TERRA COGNITA

TERRA COGNITA

The
MENTAL DISCOVERY
of
AMERICA

Eviatar Zerubavel

Rutgers University Press
New Brunswick, New Jersey

Library of Congress Cataloging-in-Publication Data
Zerubavel, Eviatar.
    Terra cognita : the mental discovery of America / Eviatar Zerubavel.
       p.   cm.
    Includes bibliographical references and index.
    ISBN 0-8135-1897-0 (cloth) — ISBN 0-8135-1898-9 (pbk.)
    1. America—Discovery and exploration.   2. Geography—
History. I. Title.
F121.Z47   1992
910′.9—dc20   92-5563
                                                  CIP

British Cataloging-in-Publication information available

To Yael
My companion on
this long voyage of
discovery called life

CONTENTS

PREFACE

I first came to America when I was seven, on a ship going from Genoa, the hometown of the man commonly hailed as its discoverer, to Colombia, the country actually named for him. Thirty-six years later I still remember quite vividly the thrill of catching my first glimpse of land following a two-week voyage across the Atlantic.

Back then, like most children, I was still fascinated by the conventional, storybook account of Columbus's heroic "discovery" of America. By the time I was sixteen, however, I found out that it had already been discovered several times before. Deeply intrigued by the idea that he was probably not "the first," I began to devour books about Columbus's precursors.

It was only a couple of years ago, however, that I discovered that even more intriguing than the story of the pre-Columbian discovery of America was the story of its *post*-Columbian discovery by Europe. It was a collection of early-sixteenth-century maps of this continent that first made me realize how very slowly it actually emerged from the ocean. That these maps in fact captured the actual formation of the entity "America" in Europe's mind immediately caught my attention, as I was just about to complete a general study of the way we construct mental entities (*The Fine Line*).

Realizing how very little actually happened in 1492 was quite a shock to me, especially given the widely publicized preparations around the world for the five-hundredth anniversary of Columbus's historic landfall in the Bahamas. Despite the heated political debate between those who glorify and those who denounce Columbus, I was struck to see how both camps were clearly overlooking the fact that what he had actually "discovered" in 1492 was a far cry from what we now call America. It was this glaring epistemological void in the current discourse about the European discovery of America that prompted me to publish this book before its five-hundredth anniversary was over.

From a very early age I have always been attracted to and intrigued by maps and timetables, those remarkable instruments we have invented to help us infuse our temporal and spatial environments with some meaningful structure. My deep interest in the structuring of time has since generated several books about schedules and calendars, yet this is my first attempt ever to also give some expression to my lifelong fascination with the structuring of space, which had very early manifested itself in an obsession with mapping. (By the time I was ten, I had already charted all the streets and bus routes of my hometown, Tel Aviv.) Studying the actual phenomenon of mapping has thus been a particularly gratifying personal experience for me.

As I was writing this book, I benefited tremendously from the great support and encouragement of Marlie Wasserman, Ken Arnold, and Barbara Kopel at Rutgers University Press as well as from the very useful comments of Richard Williams and Karen Cerulo on an early draft of the manuscript. Thanks are also due Sue Llewellyn, for a fantastic job of editing the manuscript. I am thankful for the loving environment provided by my children, Noga and Noam, and my wife, Yael, who first encouraged me to write this book and also inspired me by her own work on the social construction of historical narratives. It is to her, whom I love to discover again and again, that I dedicate this study of the mental discovery of America.

ACKNOWLEDGMENTS

Jacket, Plates 16 and 18: Biblioteca Medicea Lauren-
ziana, Florence

Plates 1, 8, and 19: Bibliothèque Nationale, Paris

Plate 2: Courtesy of the John Carter Brown Library at
Brown University

Plate 3: Biblioteca Estense, Modena

Plate 4: Beinecke Rare Book and Manuscript Library,
Yale University

Plates 5 and 6: Library of Prince von Waldburg zu
Wolfegg-Waldsee

Plate 7: Museo Naval, Madrid

Plate 9: National Archives of Canada, NMC 118245

Plates 10 and 11: Universitätsbibliothek München,
Cim.74 (= 4° Math. 863)

Plate 12: Württembergische Landesbibliothek, Stutt-
gart. Sammlung Nicolai 79, fol. 36–39

Plate 13: National Maritime Museum, London

Plates 14 and 27: By permission of the British Library.
Maps C.2.cc.4 and Maps C.7.c.17

Plates 15 and 17: Courtesy of the Hispanic Society of
America

Plate 21: Biblioteca Nazionale Centrale, Florence

Plate 22: Reproduced by permission of the Huntington

Library, San Marino, California. HM 45

Plate 23: Courtesy of the James Ford Bell Library, University of Minnesota

Plate 24: Courtesy of the Newberry Library, Chicago

Plates 28 and 29: By permission of the Houghton Library, Harvard University

Plate 30: Map copyright permission: Hammond Incorporated, Maplewood, New Jersey

TERRA COGNITA

INTRODUCTION

This year America is celebrating the five-hundredth anniversary of its discovery by Christopher Columbus. Numerous books, newspaper and magazine articles, conferences, television programs, public lectures, and exhibits are drawing the world's historical attention to October 12, 1492, the day on which Columbus made his historic first landfall on this side of the Atlantic. They all feature that day as one of the most critical moments in the history of humankind—a moment which marks the beginning of the process whereby Europe came to dominate most of the rest of the world politically, economically, and culturally, spreading Western civilization all around the globe.

Meanwhile, launching a massive moral offensive, the Left has been vigorously protesting the European colonization of the Third World. With a sense of historical irony, it is using the anniversary festivities as a rare opportunity to commemorate America's all-but-forgotten native population and to denounce Columbus's 1492 voyage as a most unfortunate event that has led only to genocide, slavery, and a disastrous destruction of the environment.[1]

The current battle waging over our collective memory is exceptionally heated. Like most cultural battles, it has

clearly probed deep into our soul. Its actual timing, however, seems to indicate that, despite their fundamental dis-
agreement over its consequences, both camps nevertheless
accept October 12, 1492, as the day on which Europe discovered America.

Yet was it really? Did Christopher Columbus indeed discover America on October 12, 1492?

◆

The notion of discovery, of course, evokes the notion
of a prior void. When we say that Columbus discovered
America, we imply that there was nobody there before
him. (To appreciate the blatant asymmetry involved in the
conventional rhetoric surrounding 1492, note how no one
ever claims that the first American Indians Columbus took
back to Spain in 1493 "discovered" Europe.)

Yet, as we all know, when Columbus "discovered"
America, millions of aboriginal Americans were already
living there. It was their ancestors who in fact had discovered it thousands of years before, when they first migrated
from Siberia into Alaska. This "minor" fact, of course, has
until quite recently been omitted from the conventional
narrative of the discovery of America since it is clearly in-
consistent with the common notion that it was discovered—and that its history, therefore, actually began—only
in 1492.

In short, only from Europe's standpoint was America
discovered by Columbus. Only from an extremely narrow
Eurocentric perspective can October 12, 1492, be seen as
a beginning. One need not share the anticolonial moral
stance of the Left to appreciate the serious epistemological
problems involved in regarding 1492 as the beginning of
American history. For the millions of Native Americans
who had lived in America for thousands of years before it

was "discovered" by Europe, it certainly marked much more of an end than a beginning.

◆

Yet, when we say that America was discovered by Columbus, we exclude from its history not only its entire indigenous population but the other Europeans, Africans, or Asians who reached it before him as well. After all, even if we disregard America's original settlers, Columbus was still not the first to have encountered it. Leif Eriksson, for one, had landed on its northern shores almost five centuries earlier.

Furthermore, the assertion that "Columbus discovered America" presupposes a number of critical assumptions we usually take for granted and therefore leave unexamined. As soon as we start to question what we actually mean by "America" or "discover," however, it becomes clear that the basis for the common notion that America was discovered by Columbus in 1492 is rather shaky.

Consider, for example, how we define America. If America comprises only the mainland, of course, then it is clearly not Columbus's famous 1492 landfall on an island lying some four hundred miles off the shores of Florida but rather his rarely celebrated first encounter with the South American mainland six years later that ought to be regarded as its actual discovery. The fact that we nevertheless regard October 12, 1492, as the day on which it was discovered indicates that we perceive America as including not just the mainland but also the various islands lying off its shores. Yet if *that* is how we define America, then the credit of its actual discovery should not even go to Columbus at all but rather to Leif Eriksson's father, Erik ("the Red") Thorvaldsson, who discovered Greenland 510 years before Columbus's first encounter with the Bahamas.

An examination of what we mean by *America* shows how its different definitions may suggest different candidates for the title of its discoverer. Yet the list of potential contenders for that title becomes even longer once we also start to examine what we mean by *discover.* As we shall see, there is more than just one way of determining what constitutes a discovery. As a result there is also more than one answer to the question "Who discovered America?"

Consider, for example, the choice between sight and touch as the most critical sense through which geographical discoveries are made. Should the credit for discovering America be given to the person who got the first glimpse of it or the one who first landed on its shores? The various possible answers to this question may very well add other, somewhat different heroes to the annals of the history of the discovery of the New World.

We normally tend to give the credit for discovering a new land to the first person who "actually" landed on its shores rather than to the first one who "merely" recognized its true cosmographic nature. We therefore usually attribute the discovery of America to Columbus rather than to the person who first realized that what Columbus had discovered and believed to be part of Asia was in fact a New World.

Yet physical encounters alone are only part of the process of geographical discovery. As critical as Columbus's 1492 landfall in the Caribbean was Europe's subsequent realization of the true nature of what he had actually discovered there. It is somewhat ironic, therefore, that the man we credit with the discovery of the New World was in fact someone who, until he died, still stubbornly refused to accept the inevitable cosmographic implications of his own discovery and kept insisting—despite overwhelming evidence to the contrary—that the lands he had encountered beyond the Atlantic were in fact part of Asia.

As we celebrate the five-hundredth anniversary of Columbus's historic "discovery of America" this year, we may forget how very little was actually discovered on October 12, 1492. In fact, as we shall see, it took Europe several centuries fully to discover America.

It certainly cannot be claimed that Europe had indeed discovered America in 1492 when its actual image of it at that time was that of a few islands off the shores of China. Yet not until 1778, 286 years after Columbus's first encounter with the Bahamas, were Europeans fully convinced that what he had discovered beyond the Atlantic was a previously unknown fourth continent that was absolutely distinct and separate from the other three—a New World, so to speak.

America was not discovered on a single day by Columbus (or by anyone else, for that matter). Its discovery by Europe was a long, slow process which lasted almost three hundred years. October 12, 1492, marked the beginning of a long mental voyage that was fully completed only in the late eighteenth century.

Europe's first image of the new lands discovered by Columbus beyond the "Ocean Sea" was quite different from our current image of America. After all, his first encounter with the New World actually revealed to his contemporaries only a fraction of it. The small group of islands he discovered on his first voyage across the Atlantic— vaguely defined as "the Indies" and including only Cuba, Hispaniola, and a few small islands in the Bahamas—was a far cry from what we now call America (see Plate 30).

Yet the discovery of America involved far more than just a widening of scope. It also entailed the recognition of the interconnectedness of all its various parts. Back in 1492 no one would have suspected that Alaska and Paraguay, for example, are both parts of a single landmass.

Current accounts of the discovery of the New World usually portray all the various discoveries made by Europeans in the Western Hemisphere since 1492 as parts of one big discovery—America. But that was not at all how they were initially perceived by the explorers themselves or by their contemporaries. Only later, in fact, were they retrospectively lumped together as parts of a single narrative. During the first few decades following Columbus's initial encounter with America, the various new lands discovered by Europe beyond the Atlantic were still perceived there as somewhat independent, separate discoveries. No one, for example, could have predicted back in 1492 that this year's Columbus anniversary would be celebrated even in the state of Washington, which was reached by Europeans only 282 years after Columbus's first landfall in the Bahamas.

As we shall see, actual geographical information about the New World was collected by Europe piece by piece over a very long period of time. Constructing the full image of America we now have, therefore, involved a slow process of putting together dozens of often-disjointed pieces of information into a single cosmographic jigsaw puzzle.

◆

It also took Europe a long time to find out that America is quite distinct and separate from Asia. In fact, as we shall see, it took Europeans almost three centuries to become absolutely convinced that the New World is fully detached from the Orient.

Part of the reason it took them so long was the fact that

Columbus's initial findings in America were evidently insufficient to determine its actual cosmographic status and had to be complemented by the findings of later explorers such as Vasco Núñez de Balboa, Ferdinand Magellan, Vitus Bering, and James Cook. Another part of the delay was a result of the fact that the process of discovery presupposes a certain readiness to accept that what one discovers may require changing the way one sees the world—a readiness Columbus and many of his contemporaries did not have.

Columbus's unexpected discovery of America was a total cosmographic shock for Europe. Ever since his first landfall on the small island of Guanahaní, every single European encounter with America only made it clearer that, contrary to Columbus's claim, there was simply no way the new lands on the other side of the Atlantic could indeed be part of Asia. The obvious discrepancy between what he and other early European explorers found in America, and Europe's own image of the Orient, was quite overwhelming.

Yet if these new lands were not part of the Orient—or, indeed, of any of the three continents allowed for by fifteenth-century European cosmography—what, then, could they be? The new geographical information about America that kept pouring into Europe after Columbus's return in 1493 from his first voyage there was clearly incompatible with the traditional cosmographic dogma. Thus it not only presented his contemporaries with a tough intellectual challenge; it inevitably also posed a serious psychological threat to them. The totally unexpected discovery of a hitherto unknown continent far beyond the conventional borders of fifteenth-century world maps certainly expanded the Old World's horizons exponentially and destroyed its naive illusion of self-containment.

How did Columbus's contemporaries indeed make sense of the discovery of America? How did they reconcile the inherent incompatibility between the new geographical evidence about the shape of the world and the old cosmographic dogma? And how did they respond to the psychological threat posed by the inevitable collapse of classical cosmography?

◆

To answer these questions, we need to stay as close as possible to the perspective of Columbus's contemporaries and "enter their minds" as best as we can. It is therefore critical that we see the New World in the light of what was known about it in Europe at the time rather than in today's terms. This can be accomplished, for example, by examining letters, diaries, and geographical treatises written by Europeans in the late fifteenth and early sixteenth centuries. Given the special use of visual imagery in cosmography, however, studying the maps that were made in Europe during that period may be preferable.

The maps of the early sixteenth century offer us the best possible picture of the way America was literally envisioned by Columbus's contemporaries, vividly documenting how they actually processed the new findings from across the Atlantic. They graphically exemplify the various responses to the discovery of the New World—from stubborn attempts to deny its novelty by forcing the new evidence into the old dogma to actual construction of a new, four-continental image of the world.

Their vivid documentation of how a slowly emerging America gradually forced itself into Europe's consciousness also makes these maps the best proof that its discovery was a long process rather than a single event. In portraying the various distorted cosmographic visions of the Western Hemisphere sustained by Europeans long after

Columbus, they demonstrate quite clearly that America as we know it was certainly not discovered on October 12, 1492.

That the discovery of America was not a single event also means that Columbus's first landfall in the Bahamas was only one—and not necessarily the most important—part of the entire process. This implies, of course, that it also took more than one person to discover America.

And indeed, the history of the discovery of the New World includes more than one discoverer and a single moment of discovery. Christopher Columbus may, of course, have played a major role in that history, but he was certainly not the only one who did. As we shall see, he was in fact only one of a long list of discoverers of America that also includes Gunnbjorn Ulf-Krakuson, Bjarni Herjolfsson, Peter Martyr, John Cabot, Alonso de Ojeda, Vicente Yáñez Pinzón, Rodrigo de Bastidas, Duarte Pacheco Pereira, Alonso Álvarez de Pineda, Juan Rodríguez Cabrillo, Sebastián Vizcaíno, Mikhail Gvozdev, Vitus Jonassen Bering, Alexei Chirikov, Juan Pérez, and Juan Francisco de Bodega, to name but some of them. While most of these men were "actual" explorers like Columbus, some were "mere" thinkers whose contribution to the *mental* discovery of America was nonetheless just as critical as theirs. An early-sixteenth-century geographer, for example, had provided Europe with its first visual image of America as an island 271 years before its insularity was empirically substantiated, and a purely conjectural—though amazingly accurate—image of a narrow strait separating North America from northeastern Asia had already been depicted on a map 167 years before it was actually reached by Bering. As those who generated such highly influential—not to mention empirically accurate—cosmographic

visions of this continent, Martin Waldseemüller and Giacomo Gastaldi certainly belong on a list of its mental
discoverers.

Like any other history, that of the discovery of America
is but a narrative and, as such, can be told in many different ways.[2] As a result such a history might feature more
than just a single "hero," and though we usually portray
Christopher Columbus as the "hero" of the discovery of
the New World, we could have just as well picked Amerigo
Vespucci, Martin Waldseemüller, Vasco Núñez de Balboa,
Ferdinand Magellan, Giovanni da Verrazano, or James
Cook, for example, for that slot. As we shall see, each of
them was equally indispensable to the development of our
present image of America.

In commemorating 1492 we inevitably distort the reality
of the process through which America was actually discovered by Europe. By highlighting only Columbus's first encounter with the Bahamas (rather than Erik the Red's with
Greenland, Bjarni Herjolfsson's with Canada, Balboa's with
the Pacific, or Gvozdev's with Alaska, for example), we implicitly choose to relegate all the rest of this long process
to semioblivion. As a result we keep holding on to the
rather simplistic notion that America was discovered by
Columbus in 1492, a gross historical misconception.

As we celebrate the five-hundredth anniversary of Columbus's 1492 landfall in the Bahamas, we must remember that commemoration is a social process and that what
we preserve from our collective past always rests on some
social convention. It is inevitable, of course, that we follow certain conventions regarding what and whom we
choose to remember. We should bear in mind, however,
that that is what we are doing and that there is always more
than just one history of "the discovery of America" to be
remembered.

CHAPTER ONE

Did Columbus Discover America?

Pre-Columbian Discoveries of America

When we say that America was discovered by Columbus, we actually exclude from its history not only its entire indigenous population but many others who may have reached its shores before him as well. Though still somewhat inconclusive, the massive anthropological, archaeological, linguistic, and folkloric evidence of pre-Columbian encounters with the New World by people other than its aboriginal settlers is nonetheless quite overwhelming.

Such evidence ranges from early European reports of "Indians" with exceptionally light or dark complexions to Old World accounts of distant lands beyond the sea and Native Americans' own memories of ancestors and gods who came to this continent from across the Atlantic. It also includes striking resemblances between equivalent words in Old and New World languages, dozens of inscriptions carved on New World rocks that contain legible messages in Old World scripts, and numerous faces on pre-Columbian sculptures and stone reliefs all over America that stun puzzled observers with their unmistakably European, Semitic, African, and Oriental features.[1]

Such abundant testimony raises many doubts about our conventional choice of Christopher Columbus as the official discoverer of America because it clearly implies a long list of others who, if they indeed did reach it before 1492, would certainly be far more appropriate candidates for that title than he. Such a list would include, for example, some possible visitors to the New World from across the Pacific, such as the Buddhist monk Hoei-Shin, who in 499 allegedly traveled from China to the faraway land of Fu-Sang in the east. It also includes some possible visitors from Europe, the Near East, and Africa (Egyptians, Cretans, Phoenicians, Hebrews, Greeks, Carthaginians, Romans, Libyans, Mandingos) who may have crossed the Atlantic long before Columbus. Neither, given the relatively short distance between the British Isles and North America, should we be surprised to learn about possible pre-Columbian transatlantic expeditions launched from Ireland (Saint Brendan in the mid-sixth century as well as Irish monks four centuries later), Wales (Prince Madoc in the late twelfth century), or Scotland (Prince Henry Sinclair in the 1390s).

We also know that at the time Portuguese seamen were trying to circumnavigate Africa in an attempt to find a direct sea route to India, they were also combing the Atlantic in search of the legendary islands of Antilia, Brasil, and the Seven Cities, and it is quite possible that they had actually reached America by 1424 as well as in the 1470s and 1480s.[2] However, even if they did reach the New World, their efforts were probably kept secret by the Portuguese crown for political and economic reasons. Such a deliberate policy of secrecy may in fact explain the paucity of records we have of possible pre-Columbian crossings of the Atlantic made not only by the Portuguese (as well as by the Phoenicians and Carthaginians long before them) but also by English (and perhaps also Basque) sailors who may have discovered the rich fishing grounds of Newfoundland some time before Columbus's first voyage to America. After all, there is some indisputable documentary evidence showing that in 1480 and 1481, and then again in 1491 and 1492, a number of ships left Bristol on their way to the islands of "Brasil" and "the Seven Cities" in the western Atlantic.[3] And though we have no proof that any of them ever did reach the New World, a letter sent by the English merchant John Day to Columbus soon after John Cabot's 1497 voyage to North America[4] nonetheless claims that the land discovered by Cabot beyond the Atlantic was in fact the island of Brasil, which had already been "found and discovered in the past by the men from Bristol." The large amount of salt shipped from Bristol to "Brasil" in the 1481 expedition also suggests some knowledge of the extensive amount of fishing one could expect to be doing there.[5] In short, it is quite possible that English merchants indeed did reach America before 1492 yet, like the Phoenicians and Portuguese before them, chose to keep their discovery secret for obvious economic reasons.[6]

All these pre-Columbian voyages to the New World may or may not have actually taken place. And even if they did, none of the evidence we now have is conclusive enough to warrant a total revision of the conventional notion that it was Columbus who discovered America. However, there is at least one chapter in the history of possible pre-Columbian contacts between Europe and America that is absolutely indisputable. That is the chapter that deals with the Norse encounter with the New World some five hundred years before Columbus.

The Norse discovery of America occurred sometime

around the year 1001, when a small group of Greenlanders led by Leif Eriksson landed on the shores of Canada.
It was the first of several expeditions led by Leif and his
family across the North Atlantic to Baffin Island ("Helluland"), Labrador ("Markland"), and Newfoundland ("Vinland"), a Norse American saga that culminated in an
ambitious attempt made by Icelander Thorfinn Karlsefni
around 1010 to establish a permanent colony of some 160
men and women in Newfoundland.

If the Norsemen had any serious plans of settling the
New World, they must have given them up before 1030.
The archaeological testimony of their presence in America
suggests that it probably lasted no more than twenty-five
years.[7] As we shall soon see, however, some sporadic expeditions to Canada were probably still launched from
Greenland as late as the mid-fourteenth century.

Unlike other possible pre-Columbian voyages to America, the documentary evidence of this brief European encounter with the New World nearly five centuries before
Columbus is quite extensive. In fact, only a few decades
after the Norse attempt to colonize it, we already find the
earliest European record of America. In a section of his
1075 history of the Hamburg archbishopric (*Gesta Ham-
maburgensis ecclesia pontificum*) that deals with the history
and geography of Scandinavia, claiming that he had actually heard about it from the king of Denmark, Adam of
Bremen first mentions Vinland as an island that "had been
discovered by many."[8] It is also mentioned shortly after
that by Ari ("the Learned") Thorgilsson in his *Íslendinga-
bók* ("The Book of Icelanders"), an early history of Iceland
dating from the 1120s. There, writing about Greenland,
Iceland's first historian notes that its native inhabitants are
the same people who also inhabit Vinland. The rather casual manner in which he mentions it seems to suggest that,

by the early twelfth century, Vinland was still quite well known to Icelanders and therefore needed no further introduction. [9]

Ari's brief allusion to Vinland is the earliest record we have of the rich Icelandic tradition that helped preserve the memory of the Norse voyages to America to this day. That tradition is best exemplified by *The Greenlanders' Saga* and *Erik the Red's Saga*, which are by far the most detailed accounts we have of those remarkable early contacts between Europe and the New World. [10] As we assess the value of these two old Icelandic tales, originally written in the thirteenth century, as actual historical documents, we must remember that their account of the Norse encounter with America comes from the very same source that also describes in great detail their discovery and colonization of Greenland, about which we have some other corroborative evidence as well. [11] In other words, the stories of the Norse discoveries of Greenland and North America are but two parts of a single narrative.

Norse America is also mentioned a couple of times in the Icelandic Annals. [12] According to these chronicles of the island's history, in 1121 Bishop Erik ("Uppsi") Gnupsson of Greenland made a voyage to Vinland, and in 1347 a ship from Greenland arrived in Iceland after having been to Markland first. Another early-fourteenth-century Icelandic manuscript lists Helluland, Markland, and Vinland as lands lying south of Greenland. [13]

All this documentary evidence, of course, predates Columbus's rediscovery of America in 1492. Yet Scandinavians' memories of their ancestors' early encounters with the New World evidently persisted long after Columbus. The memory of Bishop Erik's 1121 voyage to Vinland, for example, is preserved in a Danish poem dating from 1605 (Claus Christopherson Lyskander's *Gronlandske Chronica*) [14] as well as in Thormodus Torfaeus's 1706 book *Gronlandia antiqua*.[15] The *Groenlands annál* likewise mentions both Vinland and Helluland as late as 1623.[16]

In fact, in the wake of the European rediscovery of North America in the fifteenth and sixteenth centuries, a number of Scandinavian cartographers even tried to correlate its old Norse geography explicitly with the more recent Portuguese and French discoveries there. Icelander Sigurdur Stefánsson's (1590) and Dane Hans Poulson Resen's (1605) maps of the North Atlantic, for example, are perfect illustrations of such curious endeavors.[17] Essentially preserving the memory of the early Norse encounters with the New World, both Stefánsson and Resen still designate parts of North America by their old Norse names—Vinland, Markland, Skralinge Land[18] (after the "Skraelings," as America's aboriginal inhabitants are invariably referred to in the sagas), and Helluland. At the same time, however, Stefánsson also mentions a strait separating "Vinland" from "America."[19] Resen, in fact, goes even further, actually designating those places by both their traditional Norse names and their sixteenth-century equivalents! Vinland thus also appears on his map as "Terra Corte Rialis" (as Newfoundland, rediscovered in 1501 by Gaspar Corte-Real, indeed appears on many sixteenth-century maps), right next to "New France," a common sixteenth-century designation of Canada. Markland is likewise also designated there as "Cape Labrador."

Aside from all this overwhelming documentary and cartographic evidence of the Norse encounters with America, we also have some indisputable physical proof of their presence in the New World nearly five centuries before Columbus. The most spectacular material evidence of such presence is undoubtedly the archaeological site of L'Anse-aux-Meadows excavated some thirty years ago by

Helge and Anne Stine Ingstad at the northern tip of New-foundland. Their rather extensive findings there included a spindle whorl, a smithy, cooking pits, hearths, slag (which indicates smelting), rivets, and a bronze pin, not to mention eight unmistakable house sites.[20] The carbon datings from the site all point to the early eleventh century.

Greenland, the actual home base of America's Norse explorers, has yielded some further material evidence of such early pre-Columbian voyages to the New World. Thus, for example, in an old Norse settlement there, archaeologists have unearthed several chests made of larch, a tree virtually unknown in Greenland as well as throughout Scandinavia yet quite abundant in Newfoundland and Labrador.[21] Early layers of an old Norse settlement in Greenland have likewise yielded a lump of anthracite coal, which was also virtually unknown there (as well as in Iceland and Norway) yet could very well have been found in Rhode Island, for example.[22]

A mere two-hundred-mile-wide strait separates Greenland from Baffin Island (see Plate 30). The ability of the Norsemen, well known for their maritime skills, to cross the Davis Strait or the Labrador Sea from Greenland to Canada need not, therefore, surprise us. In fact, given the extent of their commercial as well as military activity at the time, from Sicily in the south to the Volga in the east, it would be quite surprising had they not actually reached America, whether by design or inadvertently, in the wake of their successful effort to colonize Greenland by the late tenth century.

L'Anse-aux-Meadows was first excavated by the Ingstads in 1961, but the search for some "hard" evidence of the Norse presence in America five centuries before Columbus had begun long before, with the publication of Danish historian Carl Christian Rafn's *Antiquitates Americanae* in 1837.

Rumors of Norse America were first heard outside Scandinavia when Scottish historian William Robertson in 1777 and German naturalist Johann Reinhold Forster a few years later ventured to claim that America was actually discovered by Norsemen five centuries before Columbus, and even identified Newfoundland as their landing site.[23] But it was Rafn's book that essentially introduced the non-Scandinavian world to Norse America.

At a time when Europe and America were just beginning to appreciate the captivating magic of archaeology, Rafn offered his readers some fascinating speculations about the possible physical testimony to the Norse presence in the New World five centuries before Columbus. Thus, for example, he quite typically proceeded to identify an old skeleton that had recently been dug up in Fall River, Massachusetts, as Leif Eriksson's brother Thorvald, who had been killed in America by the "Skraelings," according to the sagas, more than eight hundred years earlier. By the same token, he also offered his opinion that the enigmatic inscriptions carved on the Dighton Rock a few miles north of Fall River included the signature of Thorfinn Karlsefni, and that the old stone tower in nearby Newport, Rhode Island, was actually built in the twelfth century by Bishop Erik Gnupsson![24]

Rafn's colorful historic vision obviously whetted America's appetite for further evidence of its Norse roots, and the 150 years that have passed since the publication of his *Antiquitates Americanae* have produced numerous Norse relics ranging from petroglyphs to battle-axes.[25] Unfortunately, however, many of these have been found to be forgeries, of which the infamous Kensington Stone[26] is probably the best-known example. And yet, interestingly enough, unlike most forgeries, many of these "Norse" relics have been produced for political rather than economic reasons.

The 1892 quadricentennial anniversary of Columbus's voyage to the New World offers us a glimpse into the politics of the discovery of America, as it clearly became the setting for the first public "duel" between Columbus and Leif Eriksson over the official title of actual discoverer of this continent. The highly publicized Columbian festivities culminated in the World Columbian Exposition in Chicago, one of whose featured highlights was a spectacular display of models of Columbus's ships on Lake Michigan.[27] In a bold countermove explicitly designed to remind America of its pre-Columbian Norse "origins" (and clearly inspiring similar attempts made several decades later by Norwegian Thor Heyerdahl to demonstrate the plausibility of pre-Columbian crossings of both the Atlantic and the Pacific), a precise replica of the famous Viking Gokstad ship soon left Norway on a highly publicized voyage to Newfoundland.[28] (An almost identical display was organized last year by the governments of Norway and Iceland to counter the 1992 Columbus festivities.[29]) Such a purely symbolic crossing of the Atlantic was, not surprisingly, masterminded and organized by Scandinavians along with Americans of Scandinavian descent, and the tremendous political significance of that can hardly be overstated. Their unmistakably patriotic efforts clearly echoed those of Italian Americans, who only a few months earlier had been busy raising money for the statue of Columbus—made by an Italian and towering on top of a column of marble brought especially from Italy—that was unveiled on October 12, 1892, the day marking the four-hundredth anniversary of their fellow Italian's historic

American landfall, near the southwest corner of New York's Central Park, a site that still bears the name Columbus Circle.[30]

Whether the Norse actually "discovered" America before Columbus is clearly more than a purely academic matter. After all, Europe's first encounter with the New World is perceived as a landmark event in history, and the tremendous prestige given to the person officially regarded as its discoverer clearly rubs off on the group with which he is commonly associated. That is why Scandinavians are often so protective of the old Icelandic sagas and why Italians typically find the Norse encounter with America five centuries before Columbus so threatening.

Thus, for example, it is hardly a coincidence that the first historian who ventured to speculate that Columbus had actually heard about the Norse voyages to America on a 1477 visit to Iceland was an Icelander, Finnur Magnússon[31] (just as it is hardly surprising that the strongest advocates of the theory that America was in fact discovered by the Portuguese several decades before Columbus have typically been the Portuguese themselves[32]). Nor should we be surprised to find that, especially compared to Scandinavians, Spanish and Italian scholars by and large discredit Magnússon's theory and dismiss the very possibility that Columbus's alleged visit to Iceland—first reported by his own son Ferdinand—ever took place.[33] For if Columbus did indeed know about the Norse crossings of the Atlantic, the historical significance of his own "discovery" of America would naturally be greatly diminished.

Similarly, in 1965, when Yale University Press published the infamous Vinland Map (a fabricated pre-Columbian world map showing Norse America), it was not a coincidence that the book that accompanied the map[34] was vigorously promoted by the Scandinavian Airlines System, which also arranged for large-scale displays of the map in the windows of its ticket offices throughout North America.[35] The obvious political undertones of the symbolic contest between Columbus and his Norse challengers for the official title of discoverer of this continent were also quite evident from the immediate response to the map, whose highly ballyhooed [36] publication was scheduled, quite provocatively, for the day before Columbus Day. Not only did the map's publication stir up a heated debate within the scholarly community as well as in the popular press around the world;[37] it clearly also enraged the two national groups most strongly identified with Columbus— those claiming his origins and those claiming the sponsorship of his historic voyages to America. Thus, for example, in three front-page articles, Spain's leading daily *ABC* passionately accused Yale of trying to rob Spain of its 1492 glory by promoting the "myth" of the Norse discovery of America.[38] The Italian American community was just as furious, as evident from the bitter attacks on Yale made by the Italian Historical Society and the portrayal of the Norse encounter with America by the chairman of the Chicago Columbus Day parade as nothing but a "Nordic myth."[39] Even more revealing was the fact that, in parts of the country particularly known for their large Italian American constituencies, prominent politicians, ranging from Governor Nelson Rockefeller and Senator Jacob Javits of New York to Governor Richard Hughes of New Jersey, found it politically necessary to rise to the defense of Columbus and publicly denounce Yale's decision to publish the map.[40] As John Lindsay, New York City's Republican-Liberal mayoral candidate, told a cheering crowd of Italian Americans in Brooklyn: "Saying Columbus didn't discover America is as silly as saying DiMaggio doesn't know anything about baseball, or that Toscanini and Caruso were not great musicians or that La Guardia was not a great mayor of New York."[41] The list of the celebrated figures to whom he compared Columbus could not, of course, have been more carefully chosen and clearly highlights the political content of the heated debate about the Norse discovery of America.

The controversy surrounding the Vinland Map underscores the unmistakably political nature of any attempt to establish historical beginnings. The reconstruction of the past is inevitably affected by the politics of the present,[42] and groups always have a vested interest in promoting or destroying certain images of their past. The distinctive social and political essence of the United States, for example, is significantly different whether we begin its official history with the foundation of the Jamestown colony in 1607, the arrival of the Mayflower in 1620, or the promulgation of the Declaration of Independence in 1776. By the same token, a revisionist history such as Ivan van Sertima's *They Came Before Columbus: The African Presence in Ancient America*,[43] which claims that America was actually discovered by Africans several times before 1492, is inevitably explosive politically, since it forces us to reexamine our conventional image of the history of race relations on this continent. No wonder it has become part of the canon of the new "Afrocentrist" movement in the United States.[44]

The politics of the most serious attempt to date to contest the choice of 1492 as the official beginning of American history also sheds some light on America's apparent inability fully to accept its indisputable pre-Columbian origins. Although most historians today acknowledge the Norse encounter with the New World long before Columbus, this brief chapter in its history has not yet been officially incorporated into America's collective memory.

Despite everything that they know about the Norse presence in America nearly five centuries before Columbus, Americans evidently still regard his 1492 landfall as the beginning of American history. And if America was officially born only on October 12, 1492, nothing that happened there before that date can be considered part of its history.

One need not accept the claim that there has been an actual conspiracy against any evidence of Europe's presence in the New World before Columbus[45] to appreciate the fact that America is somewhat amnesiac about its pre-Columbian European origins. The vigor with which it celebrates its 1492 "beginnings" this year, implicitly ignoring its indisputable Norse roots, clearly demonstrates the huge difference between the sheer knowledge of historical facts and their full official incorporation into the collective memory.

What Is America?

We have thus far considered if Christopher Columbus discovered America simply on the grounds of whether he was in fact the first to get there. However, the assertion that America was discovered by Columbus is quite problematic not only factually but also epistemologically; it presupposes a number of critical assumptions which we usually take for granted and therefore leave virtually unexamined. As soon as we start questioning, for example, what we actually mean by the words *America* or *discover*, it becomes quite clear that the basis for the conventional notion that America was discovered by Columbus in 1492 is in fact rather shaky.

◆

Lest we forget, the entities we call continents are not natural units, and their actual definition is therefore not so obvious as we might think. The inescapably problematic nature of any attempt to delineate them becomes quite clear when we try, for example, to carve two separate continents out of a single, continuous landmass (as when we draw the line that separates Europe from Asia) as well as when we try to determine the continental status of islands. After all, since islands are by definition detached from the mainland, the way they are incorporated into continents is always a matter of some arbitrary convention. Furthermore, since they are sometimes situated roughly midway between two different continents, their continental affiliations may also be somewhat ambiguous. There is no absolute way to determine, for example, whether Celebes or Timor in Indonesia are "in fact" part of Asia or Australia, or whether Crete and Sardinia in fact "belong" in Europe or rather in Asia and Africa, respectively. Such questions are inevitably unresolvable except by some arbitrary social convention.[46]

The simplest and least controversial way to define any continent without getting into such problems, of course, is to restrict our definition to large chunks of continuous landmass. By including in our definition of a continent only "the mainland," and excluding the various islands lying off its shores, we certainly avoid such unresolvable epistemological quagmires. Whereas Crete may indeed be considered part of both Europe and Asia (not to mention Africa), Switzerland is unmistakably European.

And yet, if we choose to define America as a large chunk of continuous landmass, then it is definitely not Columbus's 1492 landfall on an island some four hundred miles off the shores of Florida that ought to be regarded as its actual discovery but rather his first encounter with the South American mainland. What we should then commemorate, of course, ought not to be October 12, 1492, but rather August 5, 1498, the day he first landed on the Paria Peninsula in Venezuela on his third voyage to America. We should also move Columbus Day to the first week of August and wait another six years before celebrating the five-hundredth anniverary of the discovery of America in 1998.

Perhaps the credit for the actual discovery of America should not even go to Columbus at all. Instead it should go to his fellow Italian explorer John Cabot, who "discovered" North America on behalf of King Henry VII of England in June 1497, a year before Columbus reached South America. According to a letter sent from London (by Venetian merchant Lorenzo Pasqualigo) soon after his triumphant return to England, Cabot coasted North America for some three hundred leagues,[47] which indicates that he must have discovered not only Newfoundland and Cape Breton but also the mainland itself—most probably Nova Scotia and perhaps even Maine. This strong possibility is also corroborated cartographically by the famous world map made by Basque explorer Juan de la Cosa in 1500 (Plate 7),[48] the earliest surviving map of the New World, which portrays North America's Atlantic coast as essentially English. The coast is virtually filled with English flags (starting from a cape named "Cape of England"), and the adjoining sea is literally designated as the "sea discovered by the English."[49]

And yet, with the exception of Newfoundlanders, who

indeed celebrate Discovery Day (instead of Columbus Day) on the Monday nearest June 24, the date of Cabot's 1497 American landfall,[50] we by and large ignore his first encounter with America when we commemorate its discovery by Europe. Nor, for that matter, do we ever celebrate the day on which Columbus first reached South America. The day we have chosen to commemorate as America's official birthday is the day on which he first landed in the Bahamas.

The fact that we regard October 12, 1492, as the day on which "America" was discovered indicates that we clearly perceive it as including not just the mainland itself but also the various islands lying off its shores. However, if that is how we indeed define "America," then Columbus was definitely not the first European to have reached it. The credit for discovering it must instead go to the Norseman who reached Greenland more than five centuries before Columbus's first landfall in the Bahamas (as well as some twenty years before Leif Eriksson's landfall in Canada). That traditionally unsung hero would have to be either Snaebjorn Galti, who in 978 made a first unsuccessful attempt to settle the island,[51] or fellow Icelander Erik ("the Red") Thorvaldsson, Leif Eriksson's father, who rediscovered it four years later and then returned in 985 with fourteen shiploads of settlers to lay the foundations of what soon became a permanent colony of several thousand settlers that flourished until the fifteenth century.[52]

Even a cursory glance at a map of the North Atlantic (Plate 30) should suffice to convince us that, from a purely geographical standpoint, Greenland is indeed part of America, certainly no less so than Puerto Rico or the Galápagos. A mere twenty-five-mile-wide strip of water (the Nares Strait) is all that separates its northwest coast from Ellesmere Island, which is conventionally considered part of North America, and even the Davis Strait which separates it from Baffin Island is, after all, only two hundred miles wide. In terms of distance from the nearest continental mainland, Greenland is also much closer to Labrador than to, say, Norway. In short, it certainly "belongs" in America far more than in Europe, with which it is conventionally lumped for political reasons.

The fact that we do not usually include Greenland in our definition of America is therefore quite astonishing, especially considering the fact that we rarely question, for example, the distinctly "American" continental affiliation of the islands of Bermuda, which actually lie nearly seven hundred miles off the shores of North Carolina, or Saint Thomas, which is more than five hundred miles away from the Venezuelan coast. Such glaring inconsistencies clearly underscore the tremendous blinding power of social convention.

We have already seen that it is quite possible to acknowledge the indisputable Norse presence in Canada nearly five centuries before Columbus and at the same time regard his 1492 landfall as the "official" discovery of America. By the same token, it is also quite possible to notice every time we look at a map of North America how Greenland almost touches Canada yet at the same time regard its well-documented colonization by Icelanders in the tenth century as totally irrelevant to the history of the European discovery of America. After all, how we construct "continents" in our mind is basically a matter of classification, which is a highly conventionalized social process of lumping and splitting,[53] and the essentially social mental distances we visualize between what we consider to be different continents certainly affect our perception of the actual distances between them.

Our conventional perception of Greenland as part of

Europe is a perfect manifestation of how the layout of social space often distorts our ordinary perception of distance and leads us to inflate in our minds distances between points located in what are conventionally regarded as separate chunks of space.[54] Such a tendency to "stretch" distances across mental partitions often overrides the ubiquitous "law of proximity" that normally leads us to perceive things that are closer to one another as parts of a single entity. Given the mental divides we visualize among what we consider to be separate chunks of social space (neighborhoods, countries, continents), we often perceive even short distances across them as considerably greater than much longer distances between points located within what we consider to be one and the same chunk. Thus, for example, we often perceive Nice as somehow closer to Paris than to Milan, and Houston as closer to Phoenix than to Mexico City. By the same token, we also tend to inflate in our minds the rather short distance separating Greenland from Canada and perceive it as greater than the much longer distance actually separating it from Scotland or Norway. As a result, we keep lumping it in our minds with Europe and rarely include it in our definition of America.[55]

The highly problematic nature of how we define America also becomes evident when we try to understand the special significance of October 12, 1492, in the United States, which is hardly self-evident considering the fact that Christopher Columbus never even set foot there. After all, judging from the great enthusiasm with which that date is commemorated in this country this year (as well as from the fact that it is officially celebrated here every year on Columbus Day), one might think that Columbus first landed in Miami or Boston rather than somewhere in the Bahamas. Furthermore, whereas practically every school-child in the United States knows exactly what happened on October 12, 1492, few Americans attach any special significance to April 3, 1513, the day of Juan Ponce de León's historic landfall in Florida, which is conventionally regarded as the first European encounter with what now constitutes the United States. Admittedly, it is quite possible that this country was actually discovered not by Ponce de León but by Cabot (and, as some have argued, by Amerigo Vespucci as well) sixteen years before him. (Not only does the entire Atlantic Coast of North America already appear on la Cosa's map, Florida is also portrayed rather prominently on the Cantino [Plate 3], Caveri [Plate 8], Waldseemüller [Plate 6], and Stobnicza world maps, all of which predate Ponce de León's 1513 "discovery" of Florida.) But it was definitely not discovered by Columbus on October 12, 1492.

Americans' fascination with 1492 [56] should therefore not be taken for granted. After all, societies are never casual about the particular historical events they choose to preserve in their collective memories. [57] The fact that Columbus's first encounter with the Bahamas can generate such enthusiasm in Denver and Los Angeles certainly tells us quite a lot about the distinctively "American" identity of the United States. No wonder we use the word *America* to refer not only to this continent as a whole but also to the United States in particular (as, for example, when we talk about American cars, American films, or Americans in general). In commemorating Columbus's landfall in the Bahamas, this country makes an explicit statement about its identity as part of a single "American" continent.

What Is a Discovery?

It should be quite clear by now that it was not only Christopher Columbus who discovered America. We have seen, for example, that, even if we disregard its aboriginal settlers, he was still not the first one to have encountered it. Leif Eriksson, for one, had landed on its shores long before him. A reexamination of the way we define America has brought to the surface some others (John Cabot, Erik the Red) who discovered it before Columbus.

If we examine what we actually mean by "America," we see that there is more than just one answer to the question "Who discovered America?" Yet the list of other contenders for the title of its discoverer becomes even longer once we also try to clarify what we actually mean by *discover*. As we shall now see, there is more than one way of determining what constitutes a discovery.

The first European encounter with America's west coast in 1513 offers us a rare example of the priority of seeing over touching as the critical criterion for determining what constitutes a geographical discovery. The leader of that expedition, Vasco Núñez de Balboa, in fact made a point of climbing alone to the top of the mountain from which one could see the ocean.[58] He obviously wanted to be the first European to get an eyeful of the "other sea" he had heard about from his Panamanian informants. As we all know, history has been extremely kind to Balboa. At the same time, it has all but forgotten Alonso Martín, one of the scouts he sent ahead to find the best route from the mountain to the sea below, and who in fact became the first European to reach the shore and actually touch the Pacific a day or two before Balboa himself got there and officially took possession of it for the Spanish crown.[59]

The case of Balboa and Martín, however, is rather exceptional. (Aside from the fact that Balboa was also the official leader of the expedition, we do not even have a special word for acts such as Martín's, which would have been the cultural equivalent of landfall.) History generally tends to remember and honor those who were the first ones to make an actual landfall and all but ignores those who were the first ones to cast an eye upon new lands.

The history of the Norse discoveries of both Greenland and Canada offers some excellent examples of this general pattern. Greenland, after all, was first sighted by Icelander Gunnbjorn Ulf-Krakuson several decades before Erik the Red first landed on its shores,[60] whereas Canada was encountered for the first time not by Erik's son Leif but rather by Bjarni Herjolfsson, an Icelander who in 986 quite accidentally got a glimpse of both Labrador and Baffin Island when his ship, on its way from Iceland to Greenland, was blown by the wind and carried by the current way off its course.[61] Both Gunnbjorn and Bjarni evidently decided not to go ashore and continued instead to Iceland and Greenland, respectively. As a result they clearly paid a heavy price in terms of their historical reputations. In marked contrast to those who followed in their footsteps yet also ventured to make actual landfalls, they never became real heroes. (Bjarni was belittled by Leif himself and is still criticized by modern scholars for not being a more curious explorer and going ashore.[62]) Whereas Erik and Leif have both been inducted into history's hall of fame of celebrated explorers, few people today even recognize the names of Gunnbjorn and Bjarni.

Yet the story of the "actual" discovery of Greenland by

Erik the Red is virtually inseparable from, and would be incomplete without, the story of its earlier visual discovery by Gunnbjorn Ulf-Krakuson. Erik, after all, sailed to Greenland in 982 specifically in search of the land that had been sighted by Gunnbjorn before him.[63] One cannot accept his discovery of this island yet ignore Gunnbjorn's, since both are documented in the very same sources (*The Greenlanders' Saga* and *Erik the Red's Saga*) and textually tied together into a single narrative. The only reason we have preserved the memory of Erik's discovery rather than Gunnbjorn's is our obvious cultural bias toward valuing "actual" landings over "mere" sightings as the critical component of the act of discovering new lands.

The story of Leif Eriksson's "actual" discovery of North America around 1001 is likewise virtually inseparable from the story of its visual discovery by Bjarni Herjolfsson some fifteen years earlier. Both are parts of one and the same narrative (*The Greenlanders' Saga*) and are textually intertwined in such a way that we cannot accept Leif as the discoverer of America without also recognizing Bjarni as an even earlier one. After all, according to the saga,[64] Leif sailed to America in Bjarni's own ship, which he personally bought from him, essentially reversing the former's original course. In fact, he sailed there in search of the lands Bjarni had discovered fifteen years earlier, and upon his landing in Helluland he even referred to it explicitly as one of the lands that had been originally sighted by Bjarni. His entire voyage to America was, in short, inspired by the fact that it had already been discovered by Bjarni before him. Without Bjarni's earlier visual discovery, Leif's historic "tactile" discovery of the New World might not have taken place.

There is a somewhat similar twist to the story of the European rediscovery of America in 1492. While we certainly all remember and honor Christopher Columbus, who, aside from being the leader of the expedition, also made a point of being the first person to actually set foot on the island of Guanahaní, very few of us even recognize the name of Juan Rodríguez Bermejo (also known as Rodrigo de Triana), a sailor on the *Pinta* who was actually the first to sight land in the early hours of October 12, 1492. In fact, being somewhat sensitive to the visual aspect of the act of discovering new lands (not to mention the generous 10,000-maravedí annuity promised by Ferdinand and Isabella to the first person who would *see* land beyond the Atlantic), Columbus managed to rob Bermejo of that honor as well by claiming that it had in fact been he himself who first saw a faint light coming from the island several hours earlier that evening![65]

Unlike the Norse discoveries of Greenland and Canada, the first sighting and the first landing on the island of Guanahaní actually occurred not several years or even decades apart but rather on the very same day. As a result, even had we given the credit of discovering America to Bermejo instead of Columbus, we would still be celebrating October 12, 1492, as the day on which it was discovered. Nevertheless, needless to say, the person we have evidently chosen to honor and remember was not Bermejo but Columbus, and we certainly call the day on which we commemorate the discovery of America Columbus Day rather than Bermejo Day.

And yet, though we celebrate Columbus Day and engage periodically in unmistakably "Columbian" commemorative extravaganzas such as the 1892 and current centennial anniversaries of Columbus's first voyage across the Atlantic, we nevertheless do not call this continent "Columbia"[66]

but rather "America," after Florentine explorer Amerigo Vespucci. Despite the fact that the person who actually named the New World after Vespucci later came to regret his own decision and even gave it up altogether, the name he originally chose for it has persisted for the past 485 years.

Vespucci has long been the target of bitter attacks by Columbus loyalists who consider him an impostor.[67] After all, they claim, it was Columbus and not Vespucci who "really" discovered America. Yet naming it after Vespucci was in fact a much more appropriate choice than naming it after Columbus would have been.[68] To appreciate that, however, we need to clarify further what we actually mean by "discovery."

As products of a highly action-oriented civilization, we often value "actual" physical accomplishments far more than "merely" intellectual ones. Thus, for example, we have typically named the strait which separates America from Asia not after Giacomo Gastaldi, the first European who speculated that it existed, but rather after Vitus Bering, traditionally recognized as the first one to cross it. By the same token, we normally attribute the discovery of America to Columbus rather than to the person who first realized that what Columbus had discovered beyond the Atlantic and believed to be part of Asia was in fact a "New World."

Yet physical encounters in themselves do not exhaust the highly complex process of geographical discovery, since they must also be complemented by the mental act of understanding the identity of what is being discovered. Columbus's 1492 landfall was clearly only one—and not necessarily the most important—part of the long process of discovering America. Just as critical was Europe's subsequent realization of the true cosmographic nature of what he had actually "discovered."

America is both a physical and a mental entity, and the full history of its "discovery" should therefore be the history of its physical *as well as mental* discovery.[69] Christopher Columbus was certainly one of the great heroes of the first part of this story but not the second. As we shall now see, the unsung heroes of that other, traditionally neglected part of the history of the discovery of America were the many others—"actual" sailors as well as "mere" armchair explorers—who together helped Europe to appreciate the full cosmographic significance of what he had stumbled upon on October 12, 1492.

CHAPTER TWO

The Mental Discovery of America

In March 1493, while making a brief stop at Lisbon on his way back from America to Spain, Christopher Columbus sent Ferdinand and Isabella a letter (personally addressed to Luis de Santangel and Gabriel Sánchez, two of his main supporters at the Spanish court) announcing his return from the historic voyage he had just completed across the Atlantic. A printed version of that letter, which was written in Spanish, appeared in Barcelona just a few weeks later, and the word about the new lands discovered by the Genoese sailor beyond the "Ocean Sea" soon started to spread outside the Spanish court. By the end of that year, with the help of the newly developed printing press, between nine and twelve editions of Columbus's initial account of his discoveries in the western Atlantic were already circulating throughout the continent. Aside from the original Spanish edition, several Latin editions were also printed in Rome, Antwerp, Basel, and Paris as well as a couple of Italian translations in Florence and Rome, and a German edition may have appeared in Ulm. By 1497, less than five years after his return from his first voyage to the Caribbean, eighteen editions of Columbus's *Letter to Santangel* had already been published. By the end of the fifteenth century, some ten thousand copies of his initial account of America were circulating throughout Europe.[1]

That was how Europeans first learned about the New World that had just been discovered by Columbus beyond the Atlantic. Their first image of America was based entirely on the way it was originally portrayed by him in that historic letter.

As we all know, however, that first image was quite different from our current image of the New World.[2] The small group of islands which Columbus claimed to have discovered just off the eastern shores of Asia in 1492 was a far cry from what we now call America.

Columbus's first encounter with the New World actually revealed to his contemporaries only a fractional part of this continent; the full picture of it we now have was not available to anyone in Europe at the time. The geographical information about the new discoveries beyond the Atlantic was collected by Europe only fragmentarily, piece by piece, over a very long period of time. As a result, it took Europeans many years to assimilate those pieces properly.

The construction of a full picture of America also presupposed a major cognitive adjustment on the part of Europe. It certainly cannot be claimed that Europe had discovered America in 1492, when its actual image of it then was that of a small group of islands off the coast of China! For Europeans fully to discover America, they first had to realize that it is indeed an entity that is absolutely distinct and separate from the Old World—a New World, so to speak.

The discovery of America was not a single event that took place on a single day. Rather, it was a long process that actually lasted almost three hundred years. Even if we choose to ignore its early discovery by its aboriginal settlers as well as its rediscovery by the Norse ten centuries ago, and to play up the historic significance of October 12, 1492, we must nevertheless realize that that date marks only the beginning of a long mental voyage that was fully completed only in the late eighteenth century.

In tracing the evolution of our current image of the New World, I shall focus specifically on two main aspects of the mental discovery of America, namely how it came to be seen as a *single* as well as a *separate* geographical entity that is fully detached from Asia.

A Single Geographical Entity

Sometime after 1492, America clearly came to be seen as a single geographical entity. Yet the lumping together of everything we now consider to be part of America into a single geographical entity certainly did not occur overnight. It was a long process that was far from being over when Christopher Columbus first announced his discovery to Europe back in 1493.

First, it took Europe a long time to realize the sheer scope of what Columbus had actually discovered beyond the Atlantic. Many years passed before the entire landmass we now call America fully emerged out of what was initially only a small group of islands in the Caribbean. (In fact, given his traditional reputation as the one who discovered America, we often forget that what Columbus had actually encountered during his four voyages across the Atlantic between 1492 and 1504 were only the West Indies, a narrow strip of the Venezuelan coast, and the southeast shore of Central America from Honduras to Panama. A brief glance at a map of the New World [see Plate 30] ought to remind us what a very small part of what we now call America that is.) Back on October 12, 1492, no one could yet suspect that the small island of Guanahaní in the Bahamas is in fact part of a huge continent. It took at least another two and a half centuries before a map of a single landmass that includes both Bolivia and Alaska, for example, could finally be etched in Europe's mind.

Yet the mental construction of "America" involved more than just a widening of scope. It also entailed recognizing the interconnectedness of all the various parts of this continent. When Christopher Columbus first landed in the Bahamas in 1492, it was yet unclear to anyone in Europe (or in America, for that matter) that British Columbia and Patagonia, for example, are both parts of a single landmass!

Though current accounts of the history of the discovery of the New World usually portray the various discoveries made by Europeans in the Western Hemisphere since 1492 as parts of one big discovery, that was not at all how they were perceived by the explorers themselves or their contemporaries at the time they were made. Only in retrospect were they lumped together as interconnected parts of a single narrative. When Pedro Álvares Cabral, for example, first landed in Brazil in 1500, it was not immediately apparent to everyone in Europe that the land he had just discovered south of the equator was part of the same continent that also included the islands of Cuba and Jamaica discovered a few years earlier by Columbus in the Caribbean.

During the first couple of decades following Columbus's first encounter with the New World, the various lands discovered by Europeans beyond the Atlantic were very often perceived in Europe as somewhat independent, separate discoveries. When Cabot, for example, returned to England from North America in 1497, neither of the first two reports about his voyage[3] even made a connection between the land he discovered there and those that had already been discovered in the Caribbean by Columbus. By the same token, whereas the discovery of the West Indies was generally attributed during the early sixteenth century to Columbus, that of Brazil was attributed to Cabral, while Newfoundland was even named after Gaspar Corte-Real, the Portuguese explorer who rediscovered it in 1501. Hardly anyone in Europe could have suspected back in 1501 that those three totally independent discoveries would one day be lumped together into a single entity that would be perceived as having been discovered by Columbus on October 12, 1492. Nor could anyone have anticipated back then that 1992 would be celebrated as a quincentennial anniversary even in Mexico, which was actually "discovered" by Europe a quarter of a century after Columbus's historic landfall in the Caribbean.

That the early European voyages to the New World did not produce an image of a single landmass right away is quite evident from the three world maps made between 1506 and 1508 by Giovanni Matteo Contarini (Plate 14), Johannes Ruysch (Plate 2), and Francesco Rosselli (Plate 13). These maps are clear products of several different expeditions whose findings were never fully integrated cosmographically, and they therefore demonstrate the piecemeal manner in which Europe was initially trying to put together from the discrete pieces of information brought back home by those expeditions a full picture of America. Despite their deceptively harmonious looks, they are awkward assemblages of uncoordinated fragments that clearly do not constitute a coherent whole. Nowhere is this more evident than in the way they portray North and South America as virtually unconnected to each other.

It actually took Europe much longer than that to be-

come fully convinced that the two were indeed joined together by a land bridge. Thus, for example, they still
appear as separated from each other by water on the Paris
(c. 1515) and Johann Schöner (1520 [Plate 9]) globes as
well as on the Louis Boulengier (c. 1514) and Liechtenstein
(c. 1518) globe gores and the Henricus Glareanus (c. 1510
[Plate 10]), Peter Apian (1520), and Franciscus Monachus
(c. 1527) world maps. Central Europe, it seems, was particularly reluctant to accept the fact that they were actually
connected, and an imaginary strait still separates them
from each other on the Simon Grynaeus (1532), Joachim
von Watte (1534), and Johann Honter (1546) world maps
made in Switzerland a couple of decades later. (Honter's
map, in fact, was reprinted as late as 1595![4])

Prior to the opening of the Panama Canal in 1913, North
and South America were of course joined together by a
land bridge, yet one that was even narrower than the one
which connected Africa and Asia before the construction
of the Suez Canal. Considering the fact that we regard Africa and Asia (not to mention Europe and Asia) as two entirely distinct continents, there is therefore no reason why
we should not do the same with North and South America.
Indeed, especially in the United States, they are often so
regarded, just like Europe and Asia.

And yet, while we have virtually no name that desig-
nates both Asia and Africa together, and even *Eurasia* is a
name which is only rarely used, we certainly all talk about
America much more than we do about *the Americas*. Sometime in the years after 1492, we have clearly come to perceive America as a single entity.

The image of America as a single geographical entity, of
course, is the result of active mental construction. After all,

as the case of Europe and Asia clearly indicates, continents are products of a creative act of concept formation rather than one of discovering natural clusters that already exist "out there."[5] Europe did not simply come to "realize" that America is a single entity. If it is often seen today as one, it is only because it came to be defined as such by Europe.

America is a concept, and concept formation is a mental act that usually involves language, since it is mainly our ability to assign things a common label that allows us to lump them together in our minds as a single entity.[6] It is a lot easier, for example, to form a distinct mental cluster in our minds when we have a single label (*heavy metal, conservative, postimpressionist*) with which to identify it—hence the historic significance of the moment when a single common label was first used on a map to refer to the Western Hemisphere as a whole. That label, of course, was *America*. While the name itself, as we shall see later, had been used as early as 1507 to designate parts of South America, not until 1538 was it ever applied to North and Central America as well. It was the celebrated Flemish cartographer Gerardus Mercator who first placed the name *America* on the southern as well as northern parts of the New World on his 1538 world map, thereby promoting its image as a single entity in Europe's mind.

Like common labels, commonly commemorated dates of "origin" also help consolidate mental clusters such as organizations, nations, and continents. The fact that October 12, for example, is celebrated annually not just in the Bahamas obviously shows that Columbus and 1492 have been symbolically adopted by both North, Central, and South Americans, and thus helps to promote in our minds the image of America as a single entity. By the same token, this year's commemoration of the discovery of "America" throughout the Western Hemisphere certainly helps reinforce the symbolic connectedness of what were originally several quite separate discoveries. It is clearly an anniversary of a single entity celebrating its geographical as well as historic unity.

◆

Needless to say, a continuous coastline also helps promote the image of a single geographical entity. Today, of course, we all know that, prior to the opening of the Panama Canal, North and South America were indeed a single landmass. Yet long before the continuous nature of America's eastern coastline was established as a fact, it had already been postulated by some bold visionaries.

The idea that all the lands that had been discovered beyond the Atlantic since 1492 may indeed constitute a single continent was originally put forth as early as 1501 by the survivors of the Portuguese expedition led by Gaspar Corte-Real to Newfoundland and Labrador. After having coasted more than six hundred miles of a continuous shoreline and noticed many large rivers, they concluded that the land they discovered in the North Atlantic was clearly more than an island and that it was quite possibly connected to both the West Indies and Brazil (which had been discovered just a year earlier by Cabral)! Their bold speculations about the continental nature of America were documented soon after their return to Lisbon in a report sent on October 18, 1501, by Venetian ambassador Pietro Pasqualigo:

> The crew of this caravel believes that the above-mentioned land is mainland. . . . They are also of opinion that this land is connected with the Antilles, which were discovered by the sovereigns of Spain, and with the land of the Parrots [Brazil] recently found by [Portugal's] king's vessel on their

way to Calicut. To this belief they are moved in the first place, because after ranging the coast of said land for the space of 600 miles and more, they did not find it come to an end; next, because they say they have discovered many exceedingly large rivers which there enter the sea.[7]

This extremely audacious theory evidently made a strong impression on at least some of Portugal's leading explorers and cosmographers, and a few years later the continental nature of America was already presented by Duarte Pacheco Pereira in his *Esmeraldo de situ orbis* as a well-established fact. (The book was completed in 1508, but the particular section in which he presents his vision of the New World was probably written as early as 1505.[8]) Drawing on the claims made by Corte-Real's crew[9] as well as some more recent information about South America obtained from the expedition led by Gonçalo Coelho in 1501–1502, Pacheco introduces America to King Manuel I as a full-fledged continent:

[A] very large landmass with many large islands adjacent, extending 70° North of the Equator, and located beyond the greatness of the Ocean, has been discovered and navigated; this distant land . . . extends 28½° on the other side of the Equator towards the Antarctic Pole. Such is its greatness and length that on either side its end has not been seen or known.[10]

◆

The fact that it was indeed one continuous landmass, of course, clearly helped promote America's image as a single geographical entity. As Europe came to find out during the course of the next twenty years, those early Portuguese visionaries had in fact been quite right in their bold speculations back in 1501.

Ironically, the story of how Europe actually discovered America's continental nature is but the flip side of the story of its continuous failure to find a sea passage to Asia through or around it. Every disheartened explorer who returned from America without having found that elusive passage—which, with the notable exception of the discovery of the Strait of Magellan in the far south, was first negotiated by Roald Amundsen (in the far north) only in 1903–1905—was serendipitously also promoting the image of a single American continent in Europe's mind.

From the very start, that was the big irony of the European discovery of America. Ever since Columbus's first encounter with it, it was viewed by Europe more as an impediment than as an end in itself. As a result the entire story of how Europeans discovered its continental nature was actually the story of the inadvertent results of their failed attempts to avoid it on their way to Asia.

The first European explorer to have accomplished those twin feats of failing to find a westward sea passage to Asia while at the same time inadvertently promoting the image of a continental America was John Cabot. Before leaving on his second voyage to America in 1498, Cabot made specific plans to sail southward from Canada along the North American coast until he reached the tropical regions, where he expected to find Japan.[11] As evident from the la Cosa map (Plate 7), in his failed attempt to reach Asia he clearly established the continuity of most, if not all, of North America's Atlantic coastline during that voyage. (In fact, explicit orders given by Ferdinand and Isabella in 1501 to "stop the English" in Venezuela[12] suggest that he may have actually got as far as South America.) However, other than that map there is no evidence that news of that voyage ever reached Europe.

Meanwhile, following the discovery of South America by Columbus in 1498, five Spanish expeditions managed between them to establish the continuity of its entire northern coastline.[13] Much of this endeavor was, in fact, accomplished in 1499 by the first expedition, led by Alonso de Ojeda, who sailed from the Orinoco Delta westward to the Guajira Peninsula in Colombia while another part of his fleet, led by Vespucci, was at the same time touring the Amazon Delta and Guiana to the east. The following year the entire northern coast of Brazil was explored by both Vicente Yáñez Pinzón (captain of the *Niña* in 1492) and Diego de Lepe, and by 1501 Vélez de Mendoza was already touring the northern part of its eastern coast. In 1501–1502, a fifth expedition led by Rodrigo de Bastidas (along with la Cosa) proceeded westward from where Ojeda had left off two years earlier, eventually sailing as far west as southeastern Panama.

That coastline was pushed farther north toward the end of 1502, when Columbus, on his fourth and final voyage to the New World, sailed down the Central American coast from northern Honduras to the part of Panama that had already been reached by Bastidas (about whose voyage he had probably heard several months earlier in Hispaniola).[14] Thus, by late 1502, the continuity of America's eastern coastline from Brazil in the southeast to Honduras in the northwest had already been fully established by Spain.

Portugal, in the meantime, had pushed that coastline even farther south. Though the 1501–1502 expedition led by Gonçalo Coelho failed in its attempt to find a southwest passage to Asia around South America, it certainly established the southward extent of the continent far beyond what was known in Europe until then. Even if we do not accept the claim made by Vespucci, who was on it, that they actually got as far as fifty degrees south of the equator,[15] some of the maps produced in Portugal and Italy soon after their return to Europe (especially the Kunstmann II but also the Cantino [Plate 3], King-Hamy [Plate 22], and

Caveri [Plate 8]) show quite clearly that they definitely reached Uruguay and perhaps even Argentina.

"Officially," it was the voyages of Pinzón and Juan Diaz de Solís in 1508[16] and Juan de Grijalva in 1518 that pushed America's Atlantic coastline farther northward around the Yucatán Peninsula and into the Gulf of Mexico. And after Alonso Álvarez de Pineda coasted the entire northern shore of the gulf (also failing to find a sea passage to the Pacific[17]) in 1519, the gap between Mexico and Florida was also filled in Europe's mind, as evident from Pineda's own map of the region. Yet the continuity of the coastline from Mexico to the Carolinas had obviously been established by some unknown sailors long before that. After all, the distinctively shaped gulf is quite unmistakably portrayed on Niccolo Caveri's c. 1504–1505 (Plate 8) and Martin Waldseemüller's 1507 (Plate 6) world maps as well as on Schöner's 1515 globe and the Boulengier (c. 1514) and Liechtenstein (c. 1518) globe gores.[18] In fact, in a small inset above his map (Plate 6), Waldseemüller actually provided Europe with a remarkably correct image of a continuous landmass consisting of South America, Central America, and the southeastern part of North America, thereby visually articulating the concept of a single American continent for the very first time! A virtually unbroken Atlantic coastline running from Brazil to the Carolinas is also portrayed on Waldseemüller's later maps from 1513 and 1516 (the first was included in Ptolemy's *Geography* and therefore had a wide circulation as well as considerable authority) as well as on Johannes Stobnicza's and Gregor Reisch's world maps from 1512 and 1515.[19]

The southern tip of America was finally reached in 1520 by Magellan, and four years later Europe also established the continuity of its eastern coastline all the way up to Labrador in the north. As the celebrated Florentine explorer Giovanni da Verrazano reported to King Francis I of

France on his return from his historic 1524 voyage along the North American coast: "Land has been found by modern man which was unknown to the ancients, another world with respect to the one they knew, which appears to be larger than our Europe, than Africa, and almost larger than Asia. . . . *All this land or New World which we have described above is joined together.*"[20]

Here again, it was basically his failure to find a northwest passage to China that led Verrazano to establish the continuity of America's Atlantic coastline from the Carolinas to Canada. As he himself somewhat apologetically explained to the king: "My intention on this voyage was to reach Cathay and the extreme eastern coast of Asia, but I did not expect to find such an obstacle of new land as I have found; and if for some reason I did expect to find it, I estimated there would be some strait to get through to the Eastern Ocean."[21] The striking contrast between Verrazano's great cosmographic accomplishment and his obvious disappointment over his inability to find that elusive strait captures the deep irony of the inevitable relations— so evident to Europe ever since October 12, 1492—between the failure to reach Asia and the success in discovering America:

> Here was serendipity in its true sense; in search of Cathay he had found not a passage to that land but had explored, described, and mapped the long east coast of the United States and Canada. It is probable that he gave up hope of finding a passage to his "blessed shores of Cathay" at the point in his voyage where he failed to find an open sea joining the Atlantic and the Pacific in the neighborhood of northern Maine and Nova Scotia. It must have been soon afterward that he began to find alleviation of his disappointment in the reflection that he had carried through a notable exploration as the

result of which Florida in men's minds was joined to Newfoundland. He had closed the last significant gap in knowledge of the coasts which separated Patagonia from Labrador.[22]

A brief glance at the Salviati (Plate 18) and Juan (Amerigo's nephew) Vespucci (Plate 17) 1526 world maps should suffice to demonstrate the tremendous cosmographic impact of Magellan's and Verrazano's historic voyages in the early 1520s on Europe's image of America as a single entity. The continuous landmass portrayed on those maps is a far cry from the assemblage of disjointed coasts one sees, for example, on the Kunstmann III, Contarini (Plate 14), Ruysch (Plate 2), Rosselli (Plate 13), or Kunstmann IV (Plate 31) maps. In marked contrast to the latter, it has a virtually unbroken coastline running all the way down from Labrador in the far north to Patagonia in the far south. Such unmistakably post-Magellanic as well as post-Verrazanian America is, without question, a single geographical entity, just like the one envisioned by those remarkably prophetic Portuguese sailors back in 1501.

A Separate Entity

Discovering a long continuous coastline, however, does not necessarily imply the discovery of a new continent. After all, at least in theory, the western shore of the Atlantic could still be Asia. The most remarkable thing about what Christopher Columbus had actually encountered beyond the "Ocean Sea," however, was the fact that, despite his own lame attempts to identify it as such, it was definitely not Asia. In fact, the establishment of its total separateness from Asia was the main theme underlying the long history of the mental discovery of America.

Since America was originally encountered by Europeans coming from across the Atlantic, the story of the discovery of its continental nature was primarily the story of their exploration of its eastern shore. As is evident from the Salviati map, that part of the discovery of America was basically completed after the voyages of Magellan in the south (1519–1520) and Verrazano in the north (1524). However, as the Salviati map also demonstrates quite clearly, Europe's actual understanding of what exactly lay beyond that shore was quite limited at the time. With the notable exception of Central America, Europe knew virtually nothing in 1526 about America's west coast [Plate 18]. Yet it was precisely the discovery of that coast that was necessary to establish its image as a distinct entity.

While it clearly took Europeans a long time to appreciate the full continental extent of America, it took them even longer to realize that it is also quite distinct and separate from Asia. As we all know, what was initially perceived by Columbus as a small group of islands off the shores of China turned out to be a totally unsuspected "new" continent. However, it took Europe almost three centuries to become absolutely convinced that it was indeed fully detached from Asia.

Thus one cannot really talk about Europeans' image of the "New World" before it became absolutely clear to them that it was indeed fully detached from the Old. It is the fact that it is separated from other entities that provides any entity (an individual, a nation, a room, an artistic style) with a distinctive identity that sets it apart from everything else.[23] The discovery of its actual separateness from Asia was therefore indispensable for America's distinctive identity to become established in Europe's mind.

◆

Ever since he first returned from America in 1493, many in Europe regarded Columbus's "Asian interpretation" of his discoveries beyond the Atlantic with great skepticism. None of those skeptics, however, could yet prove Columbus wrong. Such proof came only in 1513, and even more conclusively in 1520–1521, in the form of the surprising discovery of a hitherto unknown, large body of water that actually separated his "Indies" from the Orient.

The story of the European discovery of "the South Sea" was a fascinating two-act drama that unfolded over the course of seven and a half years and basically revolved around two monumental discoveries. The first act featured the celebrated Spanish explorer and conquistador Vasco Núñez de Balboa. It was soon followed by a second act featuring one of the all-time superstars of the annals of discovery—Portuguese explorer Fernão de Magalhães, better known as Ferdinand Magellan.

Balboa earned his special place in history by becoming the first European ever to reach the great ocean lying beyond America. He accomplished that formidable feat by completing a relatively modest operation that Columbus had nonetheless never attempted when he discovered Panama in 1502, on his fourth and final voyage to America— he crossed the narrow (only forty-mile-wide) isthmus that separates the Atlantic and Pacific oceans from each other at the point where North and South America meet (or used to, before the opening of the Panama Canal in 1913, exactly four hundred years after Balboa's great discovery). Inspired by natives' accounts of what he soon came to call "the South Sea," Balboa set out in 1513 to cross Darién and find it. On September 25 he finally reached his destination. Standing at the top of a mountain overlooking the Gulf of San Miguel on the southern shore of Panama, he became the first European ever to see the Pacific Ocean from the east. Four days later he actually reached the shore, entered the water, and claimed it for the Spanish crown.

As Balboa himself evidently recognized when he arranged for a notary to record the names of everyone present at that historic moment,[24] the discovery of the South Sea was a truly momentous event. In terms of its historical significance, it was at least as important and consequential as Columbus's discovery of the Bahamas in 1492. September 25, 1513, definitely constitutes a watershed in the history of the mental discovery of America, as it marks the first time any European had ever seen "the other side" of the great ocean separating the New World from the Orient.

The dramatic encounter with the Pacific was a landmark in the development of Europe's image of America as a separate continent that is fully detached from Asia, as it finally provided Europeans with the first empirical proof that Columbus had been dead wrong when he identified Central America eleven years earlier as Southeast Asia. The very existence of this ocean demonstrated to Balboa's contemporaries rather conclusively that what Columbus had discovered beyond the Atlantic simply could not have been Asia. Thus Europe found out for the first time on September 25, 1513, that the New World is indeed quite distinct and separate from the Orient.

In "discovering" the Pacific, Balboa certainly showed Europe that America is definitely bounded on the west by a considerable body of water. Yet even after he had thus established that America is indeed quite distinct from Asia, no one at the time had any idea what the true dimensions of the South Sea were. As a result Europe still had no sense of the actual extent of the New World's separateness from the Orient.

It is quite easy nowadays to determine the actual distance between the meridians of any two points across the globe by comparing their respective local solar times at the very same instant. (Since the earth completes a full 360-degree rotation on its axis every twenty-four hours, local solar times always vary by four minutes for every degree of longitude. A two-hour time differential, for example, necessarily implies a thirty-degree longitudinal differential.) Such a comparison can be done quite simply by telephone, telegraph, fax, or radio. However, since those forms of instantaneous communication across long distances were invented only during the past two centuries,[25] such methods of reckoning distances across the Pacific were clearly not available in Balboa's time. Nor was there any other precise method of determining longitude before the eighteenth century.[26]

Had any of these ways of measuring distances across the Pacific been available to Balboa's contemporaries, they would certainly have known in 1513 that the shore he had just reached in southern Panama was still about 155 degrees of longitude—that is, more than ten thousand miles—away from the Moluccas, those famous "Spice Islands" west of New Guinea that had been discovered by Portuguese explorer Francisco Serrão the year before. But, as "the Far West" and "the Far East" had at that point not been integrated into a single cosmographic picture, no one could have been faulted for maintaining—even in the wake of Balboa's discovery—that what we now know to be the largest body of water anywhere is actually only a few hundred miles wide!

That, in fact, is precisely the picture one gets from look- ing at some of the maps and globes produced in Europe soon after Balboa's historic encounter with the Pacific, and which give us a clear sense of his contemporaries' cosmo- graphic vision of the relations between the New World and the Orient even after his great discovery. As is evident, for example, from the map of the East Indies made around 1519 by Lopo Homem and Pedro Reinel, or the anony- mous map of the Pacific shown on Plate 25, even the Por- tuguese were still not certain after Balboa's discovery of the South Sea that the Moluccas were any closer to Java or Sumatra than they were to America! Nor is the vision of the Pacific much different on the 1515 and 1520 (Plate 9) globes of German cosmographer Johann Schöner.

Only within this context can we fully appreciate the unique place in the history of the mental discovery of America of the famous 1519–1522 voyage of Ferdinand Magellan and Juan Sebastián del Cano around the world. Prior to that historic voyage, even Magellan's own vision of the relations between the New World and the Orient was probably quite similar to the one expressed in these maps and globes. And when he finally managed to negotiate the treacherous strait that to this day bears his name and entered the ocean which he immediately named the Paci- fic Sea, he probably still expected to reach the Moluccas within a short time. He certainly did not anticipate that it would actually take him three months of continuous sail- ing from the day he left the south Chilean coast behind him in December 1520 to the day he finally saw the island of Samar in the Philippines.[27] Yet it was precisely this rather unexpected voyage across thousands of hitherto untraveled miles of empty sea that actually demonstrated to Europe for the first time just *how* detached the New World was from the Orient.

In the history of the mental discovery of America, Magellan's rather unique form of discovery was one which we have not considered yet. In sharp contrast to Columbus, Cabot, or Cabral, for example, it was not a particularly dramatic encounter with a hitherto-unknown shore that characterizes the special role he played in that history. Few of us, in fact, even know that it was he who discovered Chile, for example. By the same token, unlike Balboa, it was not his dramatic encounter with the edge of this continent that actually got him inducted into the hall of fame of its most celebrated discoverers. What distinguished Magellan's special brand of discovery from those of Columbus or Balboa, for example, was the fact that his was a highly nondramatic nondiscovery! Every day of his voyage across the Pacific on which he did *not* reach Asia, after all, only helped further Europe's understanding of the real cosmographic essence of America.

Only on the arrival of the eighteen survivors of Magellan's expedition, led by Basque captain Juan Sebastián del Cano, back in Seville on September 9, 1522, did Europe finally learn the true dimensions of Balboa's "South Sea" and, thus, the full extent of the New World's separateness from the Orient. Their historic three-month voyage across the Pacific proved once and for all just how wrong had been the highly underestimated calculations of the size of the earth promulgated by Columbus thirty years earlier in support of his claim that the lands he discovered beyond the Atlantic were part of Asia.

To appreciate fully the special place of the first documented voyage across the Pacific in the history of the mental discovery of America, contrast the maps and globes representing Europe's cosmographic vision of America right before Magellan's historic voyage with some of the first world maps made in Spain right after del Cano's return in 1522—such as the ones presented by Emperor Charles V in 1525 and 1526 to the papal nunzios Baldassare Castiglione and Giovanni Salviati (Plate 16) and henceforth known as the Castiglioni (*sic*) and Salviati maps as well as Juan Vespucci's (Plate 15) and Diego Ribero's 1526 and 1529 world charts. In portraying the Philippines and the Moluccas on their far left side, these maps—the first ones to show Europe the new, post-Magellanic world—basically forced their viewers to look at both the New World and the Orient at once and thereby appreciate the vastness of the ocean separating them from each other. As they very clearly demonstrate, in fulfilling Columbus's old dream of reaching the East by sailing westward, Magellan actually gave the kiss of death to his image of America as an extension of Asia.

The Spanish world charts from the late 1520s also demonstrate quite clearly that Magellan's 1520–1521 voyage across the Pacific certainly thrust Europe into a major cosmographic revolution. In the course of that voyage, "east" and "west" literally met for the first time ever and could no longer be regarded as absolute concepts. The islands portrayed on the far left side of the Salviati and Ribero maps, and therefore habitually perceived as "Far West," were nonetheless the easternmost protrusions of what was literally still regarded by Europe as "the Orient"! This confusing state of affairs was further exacerbated by the fact that, while the Philippines were indeed encountered by Magellan on his way *westward* from America to Asia, nearby Indonesian islands were around the very same time being discovered by Portuguese expeditions heading *eastward* by way of the Indian Ocean.[28] In the wake of the first circumnavigation of the earth, "east" and "west" were certainly becoming relative concepts.

◆

Magellan's 1520–1521 voyage across the Pacific certainly established once and for all beyond any doubt that the southern part of America is indeed absolutely distinct and separate from Asia. But the nature of the relations between the northern part of America and the Orient still remained virtually unexplored. After all, only in 1522 did the Spanish reach Mexico's Pacific coast,[29] and whatever lay north of it was still anybody's guess. Thus, as far as its northern regions were concerned, even Magellan's historic voyage did not complete the mental discovery of an absolutely insular New World that is fully detached from the Old.

Today, of course, we know that America and Asia are indeed totally separated from each other even in the far north, but that was definitely not something of which early sixteenth-century Europeans could be so sure. Only in the late eighteenth century, in fact, did it become absolutely clear to Europe that even the northern part of America is indeed fully detached from the Orient.

From the very beginning there was an obvious asymmetry between Europe's respective cosmographic visions of the northern and southern parts of the Western Hemisphere. As we shall see later, even its most conservative cartographers had already acknowledged by 1506 the fact that South America is absolutely distinct and separate from Asia. The same could not be said about their visions of North America, however. In fact, it took Europe more than two and a half centuries after Magellan to become absolutely convinced that the New World is indeed fully detached from the Orient in the north as well.

Such a striking difference between Europe's respective cosmographic visions of the two parts of America was a consequence of the obvious asymmetry in the extent to which Europeans were able to reach the northernmost and southernmost tips of the continent, an asymmetry which is quite understandable given the actual physical layout of the Western Hemisphere relative to the equator. Whereas Magellan had had to reach only latitude fifty-four degrees south to prove that the southern part is absolutely distinct and separate from Asia, one certainly needed to go much further than that in order to be able to prove the same thing about the northern part, which, after all, stretches well beyond the Arctic Circle (see Plate 30). And, indeed, whereas Magellan had already reached the southern tip of America by 1520, no one could reach that far in the northern part of this continent for several centuries after that. The fact that all of Europe's attempts to find a northwest passage to Asia failed also made such ventures totally unnecessary from a commercial standpoint.

Only in 1533 was Baja California discovered by Ortuño Ximénez, and it took seven more years for Francisco de Ulloa to reach the San Diego area.[30] An expedition led by Juan Rodríguez Cabrillo and Bartolomé Ferrelo then ventured as far as Cape Mendocino in northern California in 1542–1543, and it is quite possible that the area near the California-Oregon border was also reached by both the English (Francis Drake) and the Spanish (Sebastián Rodríguez Cermeño) in 1579 and 1595, respectively.[31] Then, in 1603, Sebastián Vizcaíno finally reached Cape Blanco in southern Oregon,[32] and for the next 130 years that in fact remained Europe's northernmost frontier on America's Pacific coast. Not until 1732, when the Russians finally encountered Alaska for the first time, did any European venture beyond Oregon in the Northwest.

Such very limited knowledge among Europeans about America's northwestern coast helps explain, for example, how Martin Frobisher could in 1576 still mistake Baffin

Island in northeastern Canada for Asia[33] and how, as late as 1636, New Englanders were still debating whether or not their colony actually bordered on the land of the Tartars.[34] In fact, when French explorer Jean Nicolet crossed Lake Michigan and landed in Wisconsin in 1634, he was actually wrapped in a silk robe, having expected to be welcomed by Chinese dignitaries![35]

As if to complement that rather sketchy portrait of northwestern America, Europe also knew next to nothing at the time about the northeastern provinces of Asia, vaguely known as Tartary. In sharp contrast to the great amount of firsthand information obtained by the Portuguese about Southeast Asia even by Magellan's time, it took another two hundred years before even Russia began to have a clear picture of eastern Siberia. Since that region was of no commercial interest to Europe at the time, there was also no great urgency to find out more about it.

The first European to reach the strait that separates America from Asia was Russian explorer Semen Ivanov Dezhnev, who in 1648 sailed through it from the Arctic to the Pacific.[36] Asia's easternmost cape is still named after him: Mys Dežneva. And yet, despite the fact that some historians indeed rank the cosmographic significance of his voyage alongside that of Columbus,[37] it evidently had no practical effect whatsoever on Europe's perception of the relations between the northern part of America and Asia, since his report languished in obscure Siberian archives for almost a century and was actually publicized only in 1742.[38]

And as long as Europe had no conclusive evidence that Asia was in fact totally surrounded by the ocean in the northeast, nor could it be absolutely certain that America, too, was indeed completely "closed off" in the northwest! Thus, even by the early eighteenth century, no one in Europe could definitely rule out the possibility that the Old World and the New might still be connected to each other by some yet-unexplored land bridge in the far north.[39] Indeed, even in Russia there were many (including Peter the Great himself) who still believed that they were in fact joined.[40]

Peter, of course, was practically concerned with the possibility of sailing straight from the Pacific to the Arctic. He was also urged by various European scholars (including Gottfried Leibniz) to resolve the issue of whether Asia and America were indeed separated from or connected to each other once and for all.[41] Answering that question, in fact, was the main reason he sent Danish captain Vitus Jonassen Bering to the North Pacific in early 1725, explicitly instructing him "to discover . . . whether the country towards the north, of which at present we have no distinct knowledge, is a part of America, or not."[42]

Most accounts of the history of the discovery of America are confined almost exclusively to encounters involving Europeans reaching it from the east. Few of us have even heard of Andrés de Urdaneta, the Spanish Augustinian friar who in 1565—reversing the conventional trade route and crossing the Pacific from the Philippines to Mexico instead of the other way around[43]—became the first European ever to reach the New World from the west![44] Such a totally unprecedented event has, typically enough, been absolutely ignored by most chroniclers of the discovery of America.

Urdaneta's historic eastward voyage across the Pacific obviously does not carry much symbolic weight for anyone used to thinking about America as lying in the "Far West" (rather than the "Far East," which is how it is often portrayed on Chinese maps, for example). As the Salviati map (Plate 16) so nicely demonstrates, ever since Columbus first left Spain in 1492, westward has always been the overall direction of the expansion of Europe's frontier in the "Western" Hemisphere. No wonder one would not expect Urdaneta's voyage even to be considered relevant to the history of the discovery of America.

Yet Columbus and the other Italian, English, Spanish, Portuguese, and French explorers who came to the New World from the east did not discover it all by themselves. As we shall now see, it was Russian expeditions reaching it from the west that actually made it possible for Europe to complete the process of the mental discovery of America.

Bering reached the strait that separates Alaska from Siberia (and which to this day bears his name) and sailed through it on August 13, 1728,[45] yet even that did not conclusively establish the absolute separateness of the New World from the Old. After all, throughout that voyage Bering never even saw Alaska. As a result, even after his return it was still not entirely clear whether the shores of the Chukotsk Peninsula were indeed Asia's outermost limits. And not until it was absolutely sure about the "closure" of Asia could Europe also be positive about the absolute insularity of America. Indeed, as late as 1732, when the Russian Senate made its report to Empress Anna and gave Bering his official instructions for a second voyage to the North Pacific, there were some members of both the Senate and the Imperial Admiralty College who still believed that the two continents might be joined.[46]

That year, in fact, Alaska was actually encountered for

the first time by a European when Russian geodesist Mikhail Spiridonovich Gvozdev sighted Cape Prince of Wales on the Seward Peninsula in August.[47] Gvozdev, however, believed that this "Large Country" (as Alaska was then known among the native Chukchi) was just another one of the small Diomede Islands in the Bering Strait. (Previous Russian explorers had also heard from the Chukchi rumors about this "Large Country," which even appears east of the Chukotsk Peninsula on Fedor Beiton's [1710–1711], Semen Remezov's [1712–1714], and Ivan L'vov's [c. 1710–1715] maps of the region some twenty years earlier.[48]) Like Columbus when he first encountered "the Indies," Gvozdev was totally unaware that what he had actually seen was America.

Gvozdev, however, never came to play a major role in the history of the discovery of America. Since even he did not realize the American identity of the "Large Country," it is not surprising that the great cosmographic significance of his discovery was not appreciated by his contemporaries.[49] Despite the fact that he did indeed first "discover" it, it was thus Bering and his senior officer Alexei Chirikov who, on their second voyage to the North Pacific in 1741, became the first Europeans to see Alaska as well as identify it correctly as part of America.

Despite the fact that their ships actually lost contact early in the voyage, Bering and Chirikov both reached America within a day of each other. The first was Chirikov, who on July 15, 1741, sighted one of the small islands near Prince of Wales Island in the Alexander Archipelago on the southeastern part of the Gulf of Alaska. Quite independently, the following day, Bering came within sight of Mount St. Elias on the northern shore of the Gulf.[50]

Both Bering and Chirikov clearly understood the full cosmographic meaning of their historic discoveries. The entry for July 15 in the daily log of Chirikov's ship reads quite explicitly: "This must be America," and in his official report to the Imperial Admiralty College five months later Chirikov wrote: "This land was without doubt the American coast."[51] Alaska was also referred to as America in the preface to the journal of the voyage kept by Georg Wilhelm Steller, the German naturalist who accompanied Bering.[52] Steller referred to the natives throughout his journal as "Americans," a term also used in the report made by Sven Waxel, the officer in command following Bering's death, as well as in the official report of the expedition published in 1758 by Gerhard Friedrich Müller, a leading member of the Russian Imperial Academy of Sciences, who was one of its chief organizers.[53]

That Bering and Chirikov had actually reached America was immediately grasped by some European cartographers as well. Their sightings of Alaska thus appeared on a number of maps—most notably the one made by Joseph Delisle in 1750—that in fact referred to the area quite explicitly as "America."[54] Müller, too, included in his 1758 report a map of "the Discoveries made by the Russians on the North West Coast of America," which portrayed Alaska quite clearly as part of America.

And yet, though Bering and Chirikov were absolutely correct in identifying Alaska as part of America, back in 1741 it could also have turned out to be only an Asian island, like Formosa or Japan. As we have seen earlier, America's entire Pacific coast north of Oregon remained unexplored by Europe, and its full northward and westward extents were therefore still anybody's guess. Thus, though Bering and Chirikov certainly showed that Alaska was not part of the Asian mainland, it was quite unclear yet whether it was indeed part of America.

In fact, even after 1741 some skeptics still believed

that Alaska was but an island between Siberia and North America.[55] Thus, for example, on Isaac Tirion's c. 1751 map of the Arctic regions (Plate 26), a huge gap still separates it from Oregon. And while the rest of America is portrayed in pink, Alaska is still portrayed in yellow, just like Asia!

For Alaska to be fully recognized as part of America, it was first necessary to establish the continuity of America's Pacific coastline from Cape Blanco (reached by Vizcaíno in 1603) at forty-three degrees north in the south to the shore sighted by Chirikov in 1741 at fifty-five degrees thirty minutes north in the north. Only the existence of a continuous shore joining Oregon and Alaska would indisputably confirm the latter's American identity. It was thus essential to establish the actual connectedness of Alaska and Oregon to confirm America's separateness from Asia once and for all. Only in the 1770s, however, was that finally achieved.

◆

Rumors about Bering and Chirikov's discovery of Alaska first reached Spain in 1757, and in 1761 the Spanish embassy in St. Petersburg began sending home increasingly disturbing reports about further Russian ventures into North America.[56] In 1774 and 1775, understandably alarmed by the threat of Russians encroaching on his possessions, the viceroy of New Spain, Antonio María de Bucareli, finally sent two reconnaissance expeditions led by Juan Pérez and Bruno de Hezeta to find out the actual extent of the Russian penetration of the New World. Instructed by Bucareli to sail far north and then proceed southward along the coast, both expeditions came to play a major role in the mental discovery of America. In July 1774 Pérez reached Dixon Entrance north of the Queen

Charlotte Islands,[57] and in August 1775 one of Hezeta's officers, Juan Francisco de Bodega, reached Chichagof Island at fifty-eight degrees north,[58] well beyond the shore sighted by Chirikov in 1741. Proceeding southward, both Pérez and Hezeta then explored the coast from the Alaska Panhandle down to southern Oregon, about which Europe had until then known absolutely nothing. By the end of 1775, the continuity of America's Pacific coastline had already been established as far as fifty-eight degrees north.

Yet even after 1775 it was still necessary to integrate the various discoveries of Gvozdev, Chirikov, Bering, Pérez, and Bodega in Alaska into a single coherent picture of America's northwest coast. After all, even at that point it had not been fully established that the shores sighted by Pérez in the Gulf of Alaska and Gvozdev in the Bering Strait, for example, were indeed connected. It was thus left to the great English explorer James Cook to demonstrate "that between the farthest north of the Spaniards and the farthest east of the Russians lay a continental land-mass."[59]

Cook (who had evidently learned about Pérez's and Bodega's discoveries before he left England on his third voyage to the Pacific in the summer of 1776[60]) reached Oregon in March 1778 and for the next eight months scrupulously explored America's entire northwest coast until finally stopped by the ice north of the Bering Strait.[61] Only then was the American identity of Alaska finally established once and for all.[62] (As quite evident from his own log, not until he actually completed exploring Alaska's south and west coasts was Cook absolutely convinced that the entire land east of the Bering Sea was indeed America.[63]) And only then, therefore, did it become absolutely clear to Europe for the first time that America (and not just Alaska) is indeed fully detached from Asia. In demonstrating that Alaska was in fact part of America, it was Cook who finally established the absolute separateness of the New World from the Old.

Only when Alaska's American identity was thus firmly established by Cook could the possibility that Asia and America might somehow be connected be ruled out once and for all. Only then, 286 years after Columbus's landfall in the Bahamas, was the mental discovery of America by Europe fully completed!

CHAPTER THREE

The Psychology of Discovering America

In the letter he sent to the Spanish monarchs in March 1493 (from which, as we have seen, Europe first learned about his historic voyage to America), Columbus clearly identified the lands he had just discovered beyond the "Ocean Sea" as part of Asia. He referred to them several times as "the Indies" and identified their native inhabitants as "Indians" (a misnomer we nevertheless use to this day).[1] And yet, ever since he first landed on the small island of Guanahaní on October 12, 1492, every single encounter of Europe with America only made it clearer that, contrary to Columbus's own claim, there was just no way the new lands he discovered beyond the Atlantic could be part of Asia.

The obvious discrepancy between what Columbus and other early European explorers found in America and Europe's own image of Asia was quite overwhelming. Not until the first encounter of Hernán Cortés with the Aztec city of Tenochtitlán in 1519 had any of them come across anything in the New World that bore even a remote resemblance to the great empires one would have expected to find in the Orient. Instead of the highly advanced Chinese civilization with its glamorous cities and almighty Great Khan, what they encountered on the other side of the

Atlantic were only technologically backward tribal communities living in small hamlets and ruled by small-time local chieftains.

And if the nature of America's native population did not exactly fit Europe's image of the Orient, neither did its actual geography. Every new shore discovered by Columbus during his first two voyages to the Caribbean turned out to be only one more island instead of the much-anticipated mainland. And when, in 1498, on his third voyage, he finally did reach the mainland (*tierra firme*), that discovery was soon marred by the realization that it simply could not be Asia either.

Yet if the new lands encountered by Columbus on the other side of the Atlantic were not part of the Orient, then what were they?

Late-fifteenth-century European maps such as Henricus Martellus's c. 1489 world map (Plate 4) give us a pretty good idea of how Columbus's contemporaries actually envisioned the shape of the world just before he returned to Europe from his first voyage to America in 1493. One of the most distinctive characteristics of these maps is the rather pervasive sense of closure they conveyed to their users. The world of fifteenth-century Europeans (the so-called Old World), they quite clearly tell us, was made up of a single chunk of land consisting of Europe, Africa, and Asia and surrounded on all sides by the ocean. Fourteenth- and fifteenth-century Italian and Portuguese "portolan" sea charts[2] had already offered Columbus's contemporaries some early glimpses of the Atlantic well beyond the Azores (including some legendary islands such as Antilia), yet even they could not have prepared them for what was yet to come in the wake of his historic discovery of the West Indies.

Columbus's quite unexpected encounter with America on October 12, 1492, obviously disrupted this essentially placid image of the world, as it certainly made Europe aware of the surprising extent of inhabited land far beyond the traditional frame of this seemingly self-contained picture. The inevitable cognitive implications of such a totally unanticipated—and therefore unsettling—disruption clearly exceeded those usually associated with the mere addition of new information about some hitherto unknown land. In fact, they amounted to no less than a total cosmographic shock.

How did Columbus's contemporaries make sense of the new information about the shape of the earth? How did they in fact reconcile the inherent incompatibility between the new geographical evidence about America and the old cosmographic dogma? And how did they actually respond to the inevitable psychological threat posed by the collapse of classical cosmography along with the bubble of security with which it had provided Europe for so long? After all, a clear sense of closure usually entails certainty and predictability, thereby enhancing our tranquillity.[3]

The maps from that period offer us probably the best picture of how Europe indeed made sense of its early encounters with the New World, as they quite vividly document how Columbus's contemporaries actually processed the geographical information about the new lands discovered beyond the Atlantic before there was yet a new cosmography that could fully account for it.[4] Most important, they graphically document the various responses within Europe to the discovery of America, from the stubborn attempts to force the new evidence into the old dogma to the actual creation of a totally new image of the world. In so doing, they also offer us a most vivid visual chronicle of the eventual collapse of classical cosmography, an inevitable casualty of the European discovery of the New World.

Innovation

As we have already seen, it took Europe a long time fully to "discover" America. It was long after Columbus's first encounter with the island of Guanahaní on October 12, 1492, that Europeans finally accepted the fact that what he had actually discovered on the other side of the Atlantic was a hitherto-unknown, full-fledged, fourth continent that was absolutely distinct and separate from the other three—a virtually "New World."

Part of the reason it took Europe so long fully to "discover" America, of course, was the fact that, as we have seen, Columbus's initial findings were insufficient for a full realization of its actual cosmographic status and therefore had to be complemented by later findings of other explorers such as Balboa, Magellan, Bering, and Cook. Yet part of the delay was also due to the fact that the process of discovery presupposes a certain amount of mental readiness to perceive the objects one "discovers" as something that is entirely new—a readiness that Christopher Columbus obviously did not have.

The way we perceive things is generally a function of what they actually are as well as of how we mentally approach them. And since every one of us is a member of various "thought communities" with distinctive worldviews (for example, Catholics, lawyers, baby boomers, feminists), the latter is inevitably also affected by some well-established traditions of perceiving things.[5] Nowhere is this more glaringly evident than in science, where no "fact" ever exists totally independently of some particular paradigm of interpretation through which it is mentally processed by members of some specific scientific community.[6]

As a result of this, science normally advances not only through the discovery of new facts. In fact, the history of science is full of extremely important discoveries that actually involved no new factual findings at all. Apples certainly used to fall off trees long before Newton, and the content of the dreams analyzed by Freud was in no way unusual. What was so remarkable about Newton's and Freud's discoveries, of course, were the radical shifts in perception that evidently inspired them.

Great discoveries, in short, are made not only by adding new facts but also by reinterpreting old ones through the use of new "mental filters" provided by new paradigms of interpretation.[7] In other words, they are very often the result of a mental shift that involves approaching the very same object from an entirely new perspective. By merely shifting our perception, then, we get to discover things we have never noticed before in otherwise familiar objects. Scientific revolutions, therefore, are quite often the result of purely epistemological breakthroughs.

◆

The opinion that Columbus could not have possibly reached the Orient on his first voyage across the Atlantic was actually voiced in Europe soon after his return from America in 1493. Even among Europe's leading scholars and mariners, who certainly accepted the sphericity of the earth, and who therefore regarded the idea that Asia could indeed be reached by sailing westward as quite plausible, there were nonetheless many who remained skeptical about his ability to have reached it in such a short time (only thirty-three days from the Canary Islands). As early as October 1493, only a few months after Columbus's return, Peter Martyr d'Anghiera (an Italian humanist at the Spanish court, who later wrote the first history of the New

World) wrote to the Archbishop of Braga in Portugal that many in Spain did not believe that he could have actually reached "India" on such a short voyage.[8]

Not everyone who rejected Columbus's claim to have reached the Orient had an alternative interpretation of America to offer, yet some certainly did. Thus, for example, there were some who claimed that what he had actually discovered was the legendary island of Antilia[9] (which is indeed how the Antilles got their name). Peter Martyr himself was among those who, very early on, offered a non-Asian interpretation of Columbus's discoveries beyond the Atlantic. On November 1, 1493, in a letter to Cardinal Ascanio Sforza, he actually referred to Columbus as "he who discovered the New World" (*novi orbis*).[10] A year later, in a letter to Count Giovanni Borromeo in Milan, he again referred to "the New World" (*orbo novo*) discovered by Columbus.[11] It is to Peter Martyr, therefore, that we owe the first explicit identification of America as a New World.

That the lands they discovered beyond the Atlantic were something other than Asia was also realized very early on by the European explorers of North America. In fact, when John Cabot returned to England in 1497, he was rewarded by the king for having discovered a *new* island.[12] Nor did the English and Portuguese sailors who followed Cabot to North America consider it part of Asia. As the survivors of the Corte-Real expedition, for example, described it upon their return to Lisbon in 1501, it was a land "which never before was known to anyone."[13] That was evidently also the opinion of the English merchants who came to Newfoundland that year and very soon stopped even mentioning any plans to establish an Atlantic spice-trade route to the Orient (which, after all, was what Cabot originally had in mind when he first sailed westward in search of China).[14] In fact, by 1502 their destination beyond the Atlantic was generally known in England as "the New Found Land,"[15] a rather unlikely name for China. Deliberate attempts made by both the English (for example, Cabot's own son Sebastian in 1509) and the Portuguese to circumnavigate North America in an effort to find a sea passage to the Orient further indicate that they had already understood by then that it was not Asia.[16]

It would have been quite interesting to see how Cabot actually portrayed his vision of the relations between North America and China on the "solid sphere" which he evidently displayed in London upon his return from Canada in 1497.[17] Unfortunately, that globe has never been found. In fact, with the exception of the c. 1493 Laon globe, which does not even show yet the new discoveries beyond the Atlantic, not a single globe has yet been found from the period between 1492, when Martin Behaim produced the oldest surviving terrestrial sphere only months before Europe first learned from Columbus about the existence of America, and 1507, when Martin Waldseemüller made the first globe (the gores for which have been found) known actually to show it. As a result of this unfortunate gap, when we come to examine how Europe in fact developed the visual image of a New World that is absolutely distinct and separate from the Orient, we must rely exclusively on the flat maps that were made there during what was undoubtedly the most dramatic period in the history of the mental discovery of America.

Unlike their spherical counterparts, of course, maps do not force their producers to portray explicitly their vision of the relations between different parts of the world. The very fact that they are literally bounded by their edges allows their makers, if they so wish, to refrain from making any definitive statement about the relations between America and Asia, for example. Yet that does not necessarily imply that maps are indeed always silent about such matters. In fact, in marked contrast to globes, they embody an essentially discontinuous perception of space[18] underscored by the powerful visual image of apparent insularity inevitably invoked by their edges. In a quiet yet compelling manner, they thus force their users to separate in their minds seemingly discrete chunks of space from the larger geographical contexts in which they are actually embedded.

Such highly decontextualized representation of space allows mapmakers to generate some very interesting visual rhetoric. As evident, for example, from the way groups manipulate the temporal boundaries of historical narratives, the act of drawing a boundary (such as deciding what will constitute the beginning or the end of a narrative) often obfuscates the very existence of whatever lies beyond it.[19] By the same token, by placing America on the far left side of their world maps (thereby establishing a lasting cartographic convention that has in fact prevailed to this day), early-sixteenth-century mapmakers clearly helped promote in Europe's mind the absolute separateness of the New World from the Orient.

We nowadays typically take the fact that America is almost invariably placed on the far left side of our world maps for granted. Yet that need not have been the case at all had the producers of the la Cosa, Cantino, Kunstmann II, King-Hamy, Caveri, and Oliveriana (Pesaro) maps, for example, not established that cartographic tradition in the early 1500s. Lest we forget, while these maps were all produced between 1500 and 1505, only in 1522 did Europe in fact first learn from the survivors of the Magellan expedition the true dimensions of the ocean that actually separates the New World from the Orient. Given what was known in Europe about "the Indies" around 1500, it would have made at least as much sense for these visual pioneers to place them on the far right side of their world maps as on the far left. However, that was clearly not what they chose to do.

Boundaries by and large distort our ordinary perception of space (physical as well as mental). Drawing a line between any two entities basically widens the gap we envision between them,[20] thereby reinforcing their mental separateness from each other. After all, despite the fact that the Bering Strait is almost twice as narrow as the Korea Strait, because of the way we conventionally split the world in our mind into continents we rarely notice that Alaska is in fact much closer to the eastern shores of the Asian mainland than is Japan. In applying a somewhat similar logic of mental splitting, the practice of placing America at the far left side of world maps such as the Cantino (Plate 3), King-Hamy (Plate 22), and Caveri (Plate 8) likewise reinforced its mental separateness from Asia, thereby anticipating by a full decade Balboa's empirical affirmation of their actual physical separateness. It was also helped, of course, by the fact that on all these maps the east coast of Asia was clearly separated from the right edge of the map by the China Sea. By choosing to portray Asia in that manner, the producers of those maps clearly expressed their belief that it was in fact quite separate from both Cuba and Central America—in marked contrast to what Columbus expected Europe to believe.

Most of the users of those maps, of course, were well aware of the fact that the earth is round and therefore does not really end beyond the left edge of the map. Nonetheless, the dramatic visual effect created by that edge was quite powerful and certainly helped promote the mental separateness of the New World from the Orient in Europe's mind. Its placement on the far left rather than the far right side of world maps—totally separated visually from Asia—from the very beginning mentally lumped America together with (in fact, as an extension of) Europe into what later came to be known as "the West" and kept it from being perceived, as it very well could have been at least until 1522, as an extension of the Orient. (In fact, by October 1494, Peter Martyr had already placed the newly discovered lands beyond the Atlantic in what he came to call "the Western hemisphere."[21])

◆

The Cantino, Kunstmann II, King-Hamy, Caveri, and Oliveriana (Pesaro) maps were all drawn between 1502 and 1505 by Portuguese cartographers or Italians using Portuguese prototypes. That they played such a major role in the history of the mental discovery of America was hardly a coincidence. After all, it was a 1501–1502 Portuguese expedition to South America that first led Europe to consider seriously the existence of a New World that was absolutely distinct and separate from the Old. This was the expedition that was led by Gonçalo Coelho[22] yet is commonly associated with one of its other members, Florentine explorer-cosmographer Amerigo Vespucci.

When Vespucci left Portugal in 1501 on his way to South America, he already had good reason to suspect that the huge landmass discovered by Columbus in 1498—and shown since then by Ojeda, Pinzón, Lepe, Cabral, Vélez de Mendoza, and himself to stretch at least from Colombia to Brazil—was something other than Asia. By the time he returned to Europe after having reached Uruguay and perhaps even Argentina, he was quite positive that it was not Asia. The most critical piece of negative evidence, of course, was the seemingly endless southward extent of the

South American coastline, which stretched several thousand miles beyond the tropical latitudes in which one could still have expected to find the sea passage through which Venetian traveler Marco Polo had sailed two hundred years earlier from the South China Sea around the southeastern horn of Asia into the Indian Ocean. This huge southern landmass simply could not be Asia.

That was clearly the conclusion reached by Vespucci by the time he left South America in 1502. And in a letter he sent to his Florentine patron, Lorenzo de' Medici, soon after his return to Lisbon, he announced the discovery of "a new land," which he "observed to be a continent."[23]

Vespucci's new cosmographic vision of South America was soon being disseminated throughout Europe with the considerable help of the printing press. In fact, the tremendous success of his account of that continent was one of the earliest manifestations of what would soon become a new form of stardom: authorship of printed best-sellers. It was not exactly Vespucci's own account, however. In August 1504 a pirated Latin translation of his letter to de' Medici was published in Vienna under the highly suggestive title *Mundus Novus* ("New World"). Its author, using Vespucci's name, announced the discovery of a "new continent," which was unknown to the ancients and which "we may rightly call a new world."[24] The following month another pamphlet purportedly authored by Vespucci appeared in print—a letter he had allegedly written to the ruler of Florence, Piero Soderini, describing his "four voyages" to America. Both *Mundus Novus* and *Lettera al Soderini* were soon reprinted in several dozen Latin, German, Italian, Czech, and Dutch editions in Germany, France, Belgium, Italy, Bohemia, and Switzerland.[25]

Despite the fact that they were most probably forgeries,[26] both *Mundus Novus* and *Lettera al Soderini* clearly played a major role in the history of the mental discovery of

America. Because of their tremendous popularity throughout Europe[27] they were largely responsible for getting
Europeans for the first time to define at least some of the
lands discovered beyond the Atlantic as a new continent.

The cartographic evidence of Vespucci's pivotal role in
the mental discovery of America is quite overwhelming.
Soon after the publication of *Mundus Novus*, European
mapmakers indeed began to identify South America as
such. Thus, for example, it was designated "Mundus Novus" on the c. 1504–1505 Oliveriana (Pesaro) map, Johannes Ruysch's 1507 world map (Plate 2), and Francesco
Rosselli's c. 1508 oval-projection world map as well as on
the Lenox (Hunt) and Jagellonicus globes from c. 1510. In
fact, it was designated "Mondo Novo" even on a 1506
sketch-map designed by Alessandro Zorzi to embellish an
Italian translation of Columbus's letter to Ferdinand and
Isabella describing his fourth voyage to America, which his
brother Bartholomew brought to show the pope (Plate 21).

Calling the southern part of America a "new world" certainly implies an understanding that it is something quite
separate from Asia. But the way it was actually portrayed
on post-Vespucci maps and globes is an even clearer testimony to the tremendous impact of Vespucci's cosmographic vision of this continent on Europe. As we can see
on the Lenox and Jagellonicus globes, it was soon envisioned quite explicitly as a totally separate island. So compelling, in fact, was Vespucci's vision of southern America
as a New World that its image as an island was soon accepted even by Europe's most conservative cartographers.
Thus, for example, between 1506 and 1508, even Ruysch,
Rosselli, and Giovanni Matteo Contarini, whose world maps
literally portray northern America and eastern Asia as one
and the same, nonetheless pictured southern America as
absolutely distinct and separate from the Orient. Indeed,

they were so sure that it was not attached to Asia that they already portrayed its western shore long before Europe had actually reached the Pacific (Plates 2, 13, and 14). Though it took several more years before Balboa indeed affirmed their actual separation from each other by a wide ocean, Europe had clearly already accepted Vespucci's vision of a New World that is quite distinct and separate from the Old.

Despite the fact that in his letter to Lorenzo de' Medici Vespucci had explicitly identified southern America as a new land that he "observed to be a continent," his specific use of the term *continent* in that context was nevertheless quite vague and did not necessarily imply a full acknowledgment of America's cosmographic status as equivalent to those of Asia or Africa.[28] Such an acknowledgment was soon made, however, by Portuguese cosmographer Duarte Pacheco Pereira, who in 1505 claimed in his book *Esmeraldo de situ orbis* that, aside from Europe, Africa, and Asia, the earth also has "a fourth part" that was virtually unknown to the ancients.[29] By explicitly identifying America as a "fourth part of the earth," Pacheco thus went a step beyond Vespucci in establishing its cosmographic status as a full-fledged continent that stands on an equal footing with the other three (thereby leading a well-known historian of the discovery of America to propose that perhaps it ought to be renamed "Pacheca"[30]).

Furthermore, whereas Vespucci's "New World" was basically confined only to South America, Pacheco's vision of *a fourth continent* definitely included both South and North America. After all, as we have seen earlier, the "new" continent he introduced to King Manuel in his book presumably extended seventy degrees north of the equator. This "fourth part of the earth" clearly included *all* the lands that had been discovered by Europe beyond the Atlantic since 1492.

Probably as a result of Portugal's official censorship of maps and documents relating to its new discoveries,[31] despite the fact that it was written between 1505 and 1508 *Esmeraldo de situ orbis* was actually published only in the nineteenth century and was therefore originally read by only a handful of people other than King Manuel. Yet the audacious idea that the newly discovered lands beyond the Atlantic might in fact constitute a totally "new" continent was actually expressed around the very same time in a geographical treatise published in April 1507 in the small town of Saint-Dié in the Vosges Mountains, some forty miles southwest of Strasbourg.[32] The treatise, *Cosmographiae Introductio*, was written by German cosmographer Martin Waldseemüller, a member of the Gymnasium Vosagense,[33] a small group of scholars and intellectuals working in Saint-Dié under the patronage of Duke René of Lorraine. Having first mentioned Europe, Africa, and Asia, Waldseemüller then noted that "the earth is now known to be divided into four parts,"[34] further describing the fourth part as "an island, inasmuch as it is found to be surrounded on all sides by the ocean."[35] This was the first explicit statement ever made about America's insular nature (and, thus, about its absolute separateness from Asia).

Like Pacheco, Waldseemüller clearly meant to portray America as a full-fledged continent that stands on an equal footing with Europe, Africa, and Asia. He tried to accomplish that by explicitly identifying it as "a fourth part" of the earth as well as by giving it a name that would be clearly equivalent to theirs.[36] Choosing to name it after Vespucci, he thus wrote: "Inasmuch as both Europe and Asia received their names from women, I see no reason why any one should justly object to calling this part Amerige, i.e., the land of Amerigo, or America, after Amerigo, its discoverer."[37] He used that name in the book itself (which also included, aside from his own treatise, a new translation of "Vespucci's" *Lettera al Soderini*) as well as on the world map (Plate 6) and terrestrial globe that accompanied it.

Yet Waldseemüller deserves a special place of honor in the history of the mental discovery of America not just because it was he who gave it its present name. Far more critical to the development of Europe's image of America as a New World that is absolutely distinct and separate from the Old was the highly evocative way in which he visually articulated his revolutionary cosmographic theory on the map and globe that accompanied his treatise. It is to him more than anyone else, in fact, that we owe our present visual image of America.

Waldseemüller's 1507 world map (Plate 5) was the first one ever explicitly to portray America's insular nature and, thus, its absolute separateness from Asia. Neither the la Cosa, Cantino, King-Hamy, Kunstmann II, or Caveri map, after all, had ever made an unequivocal statement to the effect that it was absolutely distinct and detached from Asia by explicitly showing its western limits. The only earlier map, in fact, to have depicted America's west coast was Contarini's world map from the year before (Plate 14), but it did that only with its southern part, having essentially omitted Central America altogether and portrayed North America as an Asian promontory. It was thus Waldseemüller's 1507 world map that first provided Europe with that critical component of our present image of America as an island that is absolutely distinct and separate from Asia. The fact that it was made more than six years before any European had actually seen America's west coast, of course, makes his amazingly accurate portrayal of its general shape (especially on the small inset map above the main map [Plate 6]) all the more remarkable. In fact, in that inset map, Waldseemüller actually went one step further in consolidating America's status as a distinct continent by portraying it for the first time ever *east* of Asia, with a wide ocean between them. Whereas on the Cantino, King-Hamy, and Caveri maps its separateness from Asia was still only implied, on this map it was quite explicitly articulated for the first time.

Waldseemüller's bold new vision of the newly discovered lands beyond the Atlantic evidently received a lot of immediate attention. Several editions of the *Cosmographiae Introductio* were printed and a thousand copies of his world map sold the very year they were published,[38] and it is quite clear that they played a pivotal role in shaping Europe's image of America as a distinct continent. In visually postulating its entire west coast more than six years before any European had actually gotten even a glimpse of it, Waldseemüller clearly planted the idea of its total separateness from Asia in Europe's mind.

The cartographic evidence of Waldseemüller's tremendous influence on early-sixteenth-century Europe's vision of the New World is quite overwhelming. The name by which he chose to designate the southern part of the new continent on his own map ("America"), for example, soon appears on the 1510 world map, pre-1520 map of the Pacific (Plate 11), and pre-1520 map of the Southern Hemisphere of Henricus Loritus Glareanus,[39] as well as on the Cornelius Aurelius (1514), so-called Leonardo da Vinci (c. 1514), and Peter Apian (1520) world maps, Johann Schöner 1515 globe (which was also accompanied by a cosmographic treatise that included a special chapter on "America, the fourth part of the world"[40]), and Louis Boulengier (c. 1514) and Liechtenstein (c. 1518) globe gores. His considerable influence is also quite evident from the general shape of America on most of those maps, globes, and gores, clearly modeled after his original. By the same token, the image of America on Johannes de Stobnicza's 1512 world map is but a mere replica of the one on Waldseemüller's 1507 inset map.

Furthermore, though it actually took another 271 years before the absolute separateness of North America from Asia was conclusively demonstrated by Cook, many European cartographers even during the early part of the sixteenth century already envisioned the two as indisputably detached from each other. Despite the total lack of any empirical evidence, they nevertheless preserved on their maps and globes Waldseemüller's original image of North America as absolutely distinct and separate from northeast Asia. Consider, for instance, aside from all the aforementioned maps, globes, and gores, the Simon Grynaeus (1532), Joachim von Watte (1534), Gerardus Mercator (1538), Battista Agnese (1542), Sebastian Münster (1544 [Plate 23]), Gemma Frisius (1544), and Michele Tramezzino (1554) world maps, as well as the c. 1515 Paris globe and the Georg Hartmann (1535) and François Demongenet (1552) globe gores. They all portray America as fully detached from Asia even in the far north—an absolutely insular fourth continent totally surrounded on all sides by the ocean, just as Martin Waldseemüller first envisioned it back in 1507.

Despite Waldseemüller's tremendous influence on the way Europe came to view America, not until the late eighteenth century did it have any conclusive evidence that it was indeed fully detached from Asia even in the far north. For nearly three centuries European cartographers were basically promulgating on their globes and world maps an audacious cosmographic theory which, given the actual geographical information that was available to them, had no basis whatsoever in reality!

Yet maps and theories do not only reflect actual geographical realities. They very often portray the purely speculative, empirically unsubstantiated ideas of the people who originate them. In so doing, however, they sometimes help generate amazingly correct new cosmographic visions even when there is no evidence yet to support them. Long before his theory was indeed proved to be correct, Waldseemüller had already provided Europe with a most compelling first image of an absolutely insular America. As we shall see later, that was also true of the purely conjectural—though, prophetically enough, empirically correct—image of a narrow strait separating North America from northeast Asia generated by Venetian cosmographer Giacomo Gastaldi 167 years before Bering actually reached it.

The contribution of a "mere" armchair explorer like Waldseemüller to the mental discovery of America was thus at least as critical as that of an actual explorer like Christopher Columbus. And it was certainly not a result of his having had more factual knowledge about this continent. In fact, since he probably never saw America at all, he clearly had much less information about it than Columbus, who, after all, had spent most of the period from 1492 to 1504 there. Rather, it was his readiness to reexamine the information he did have about it from a totally new cosmographic perspective that enabled Waldseemüller to envision an absolutely insular America more than 270 years before Cook finally demonstrated that it is indeed fully detached from Asia.

And indeed, it was their readiness to look at what Columbus had originally identified as "the Indies" from a totally new cosmographic perspective and essentially redefine it as a hitherto unknown continent that set Amerigo Ves-

PLATE 1: Oronce Finé's cordiform world map, 1534. *North America is portrayed as an extension of East Asia. Indonesian islands are placed just off the shores of Mexico.*

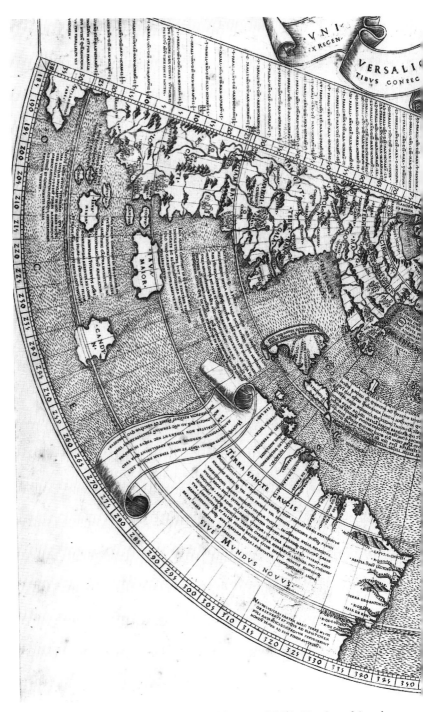

PLATE 2: Johannes Ruysch's world map, 1507. *North and South America are virtually unconnected. While the latter is designated as "Mundus Novus," the former is portrayed as an Asian peninsula.*

PLATE 3: The "Cantino" planisphere, 1502. *America on the far left side of the map is visually separated from Asia on the far right, yet both North America and Northeast Asia are cut off at the edges of the map.*

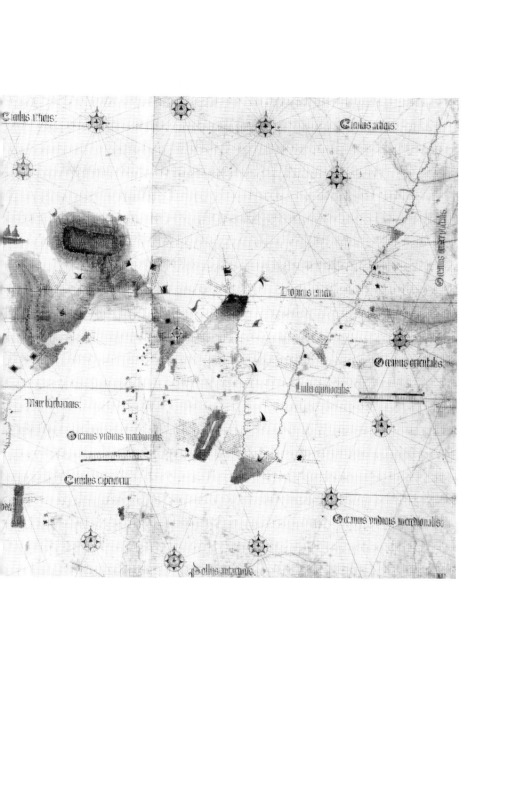

Circulus articus.

Circulus articus.

oceanus amer[...]alis

Tropicus canci

oceanus orientalis.

Linha equinocialis.

Mare barbaricus.

Oceanus yndicus meridionalis.

Circulus capricorni.

Oceanus yndicus meridionalis.

Polus antarticus.

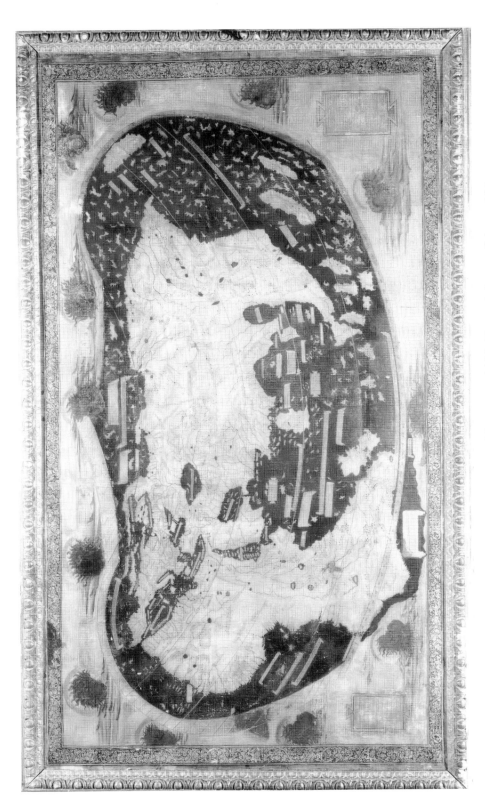

PLATE 4: Henricus Martellus's world map, c. 1489. *Europe's image of the world just before 1492. Notice the pervasive sense of closure.*

PLATE 5: Martin Waldseemüller's world map, 1507. *The first visual image of an absolutely insular America. South America is designated for the first time as "America."*

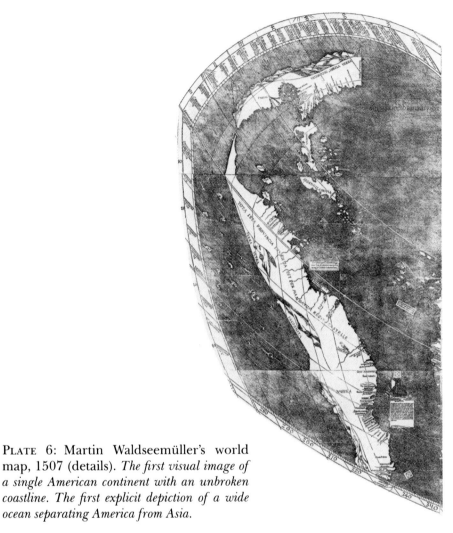

PLATE 6: Martin Waldseemüller's world map, 1507 (details). *The first visual image of a single American continent with an unbroken coastline. The first explicit depiction of a wide ocean separating America from Asia.*

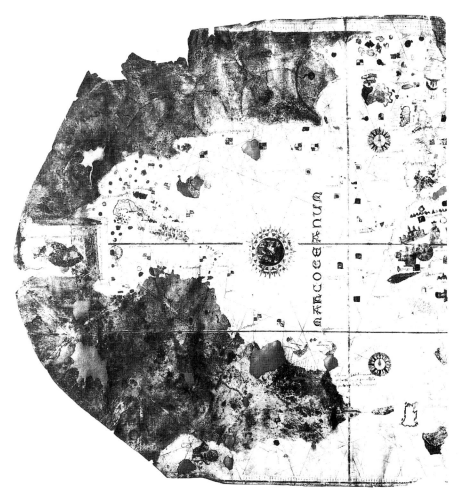

PLATE 7: Juan de la Cosa's world chart, 1500 (detail). *North and South America are both cut off at the left edge of the map. The nature of the relationship between them is also still unresolved.*

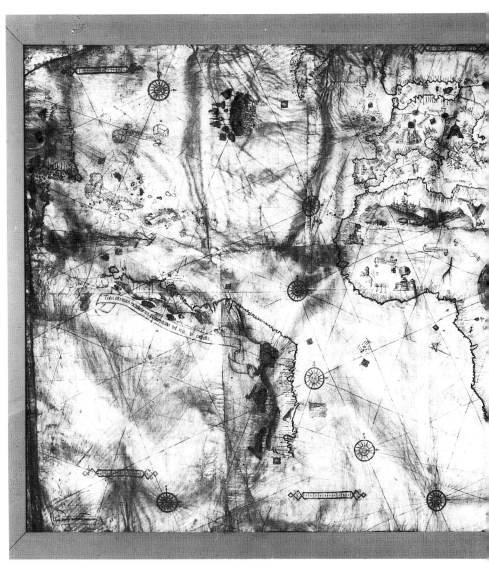

PLATE 8: Niccolo Caveri's world chart, c. 1504–1505. *America on the far left side of the map is visually separated from Asia on the far right, yet both North America and Northeast Asia are cut off at the edges of the map. Notice the Gulf of Mexico.*

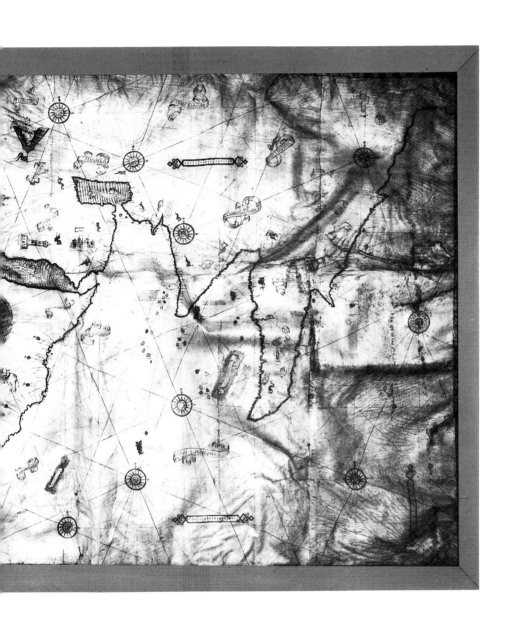

PLATE 9: Johann Schöner's globe, 1520 (facsimile of part). *America is portrayed as separate from Asia, yet the Pacific is still narrow and Japan is very close to Central America. North and South America are still portrayed as separate from each other.*

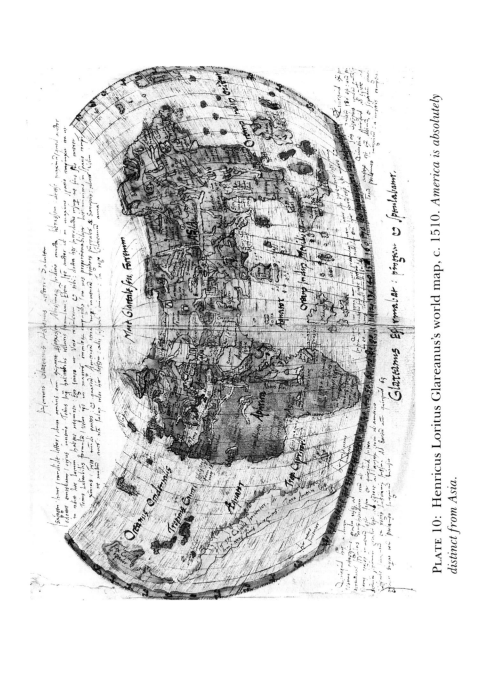

PLATE 10: Henricus Loritus Glareanus's world map, c. 1510. *America is absolutely distinct from Asia.*

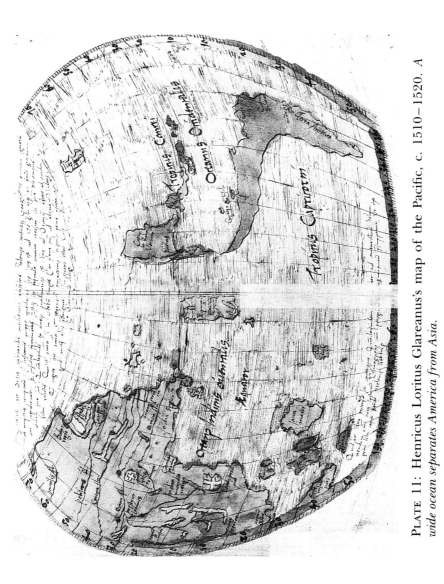

PLATE 11: Henricus Loritus Glareanus's map of the Pacific, c. 1510–1520. A
wide ocean separates America from Asia.

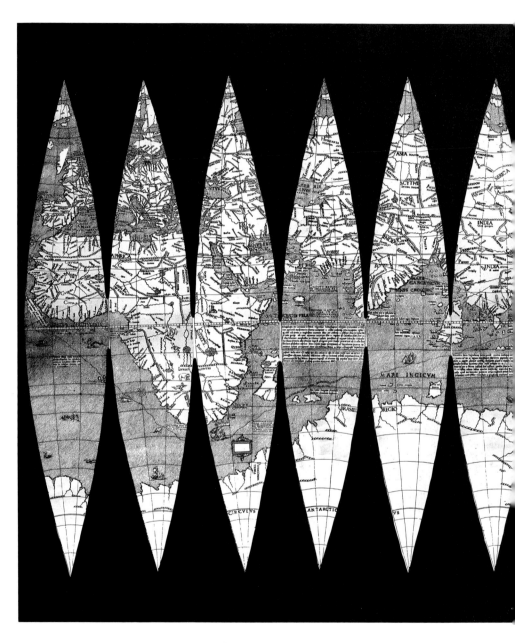

PLATE 12: Anonymous globe gores, c. 1535. *Mexico and China are portrayed as one and the same. New Spain is featured right next to Mangi and Cathay.*

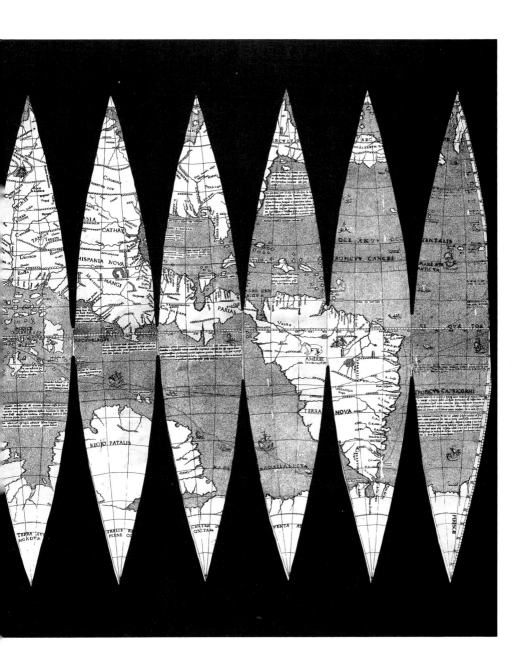

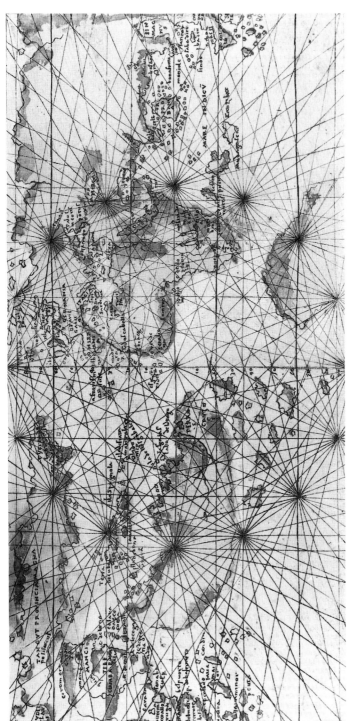

PLATE 13: Francesco Rosselli's marine chart of the world, c. 1508. *The China Sea and the Caribbean are portrayed as one and the same. Central America is placed in South-east Asia. North America is portrayed as an Asian peninsula.*

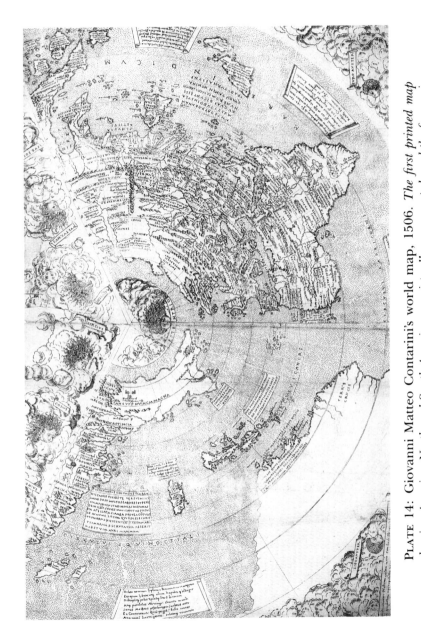

PLATE 14: Giovanni Matteo Contarini's world map, 1506. *The first printed map showing America. North and South America are virtually unconnected, and the former is portrayed as an Asian peninsula.*

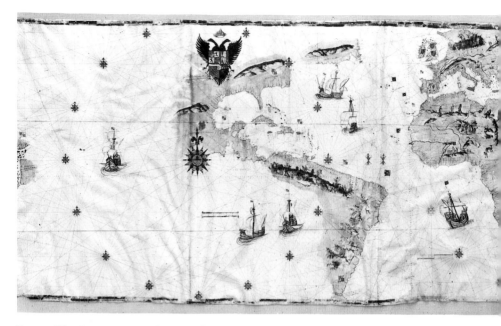

PLATE 15: Juan Vespucci's world map, 1526. *One of the first post-Magellan representations of the Pacific Ocean.*

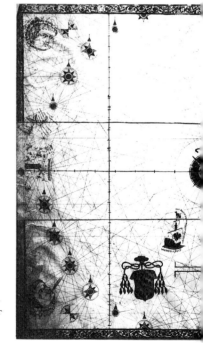

PLATE 16: The "Salviati" world map, c. 1526. *Notice how "east" and "west" meet at the far left side of the map.*

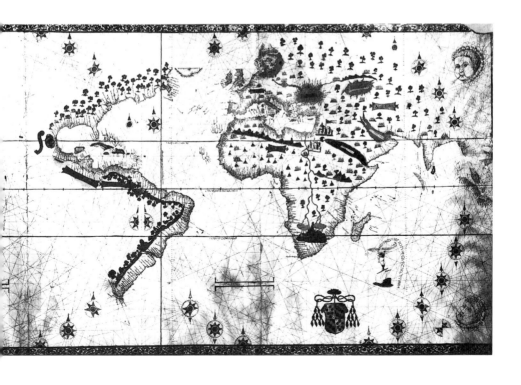

PLATE 17: Juan Vespucci's world map, 1526 (detail). *In marked contrast to the sharp delineation of its Atlantic Coast, most of America's west coast is still missing.*

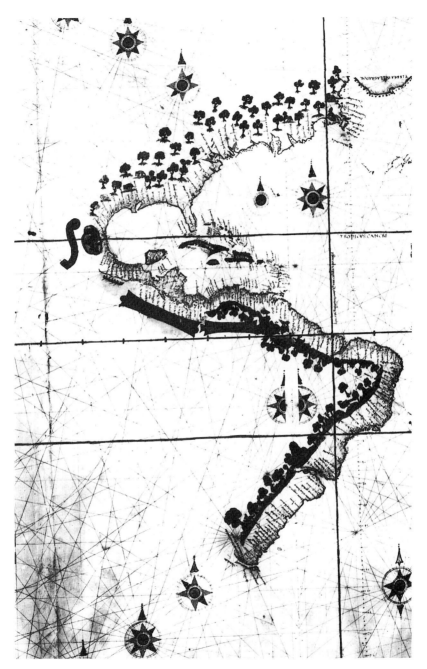

PLATE 18: The "Salviati" world map, c. 1526 (detail). *An extremely Eurocentric representation of the New World. Notice the tension between the known and the still unknown. America is still literally open-ended in the west.*

PLATE 19: The "Gilt (or De Bure)" globe, c. 1528.

PLATE 20: The "Gilt (or De Bure)" globe, c. 1528 (facsimile of part). *North America is portrayed as flowing into China. Cathay is featured right next to the Gulf of Mexico. Japan and the Yucatán are portrayed as one and the same.*

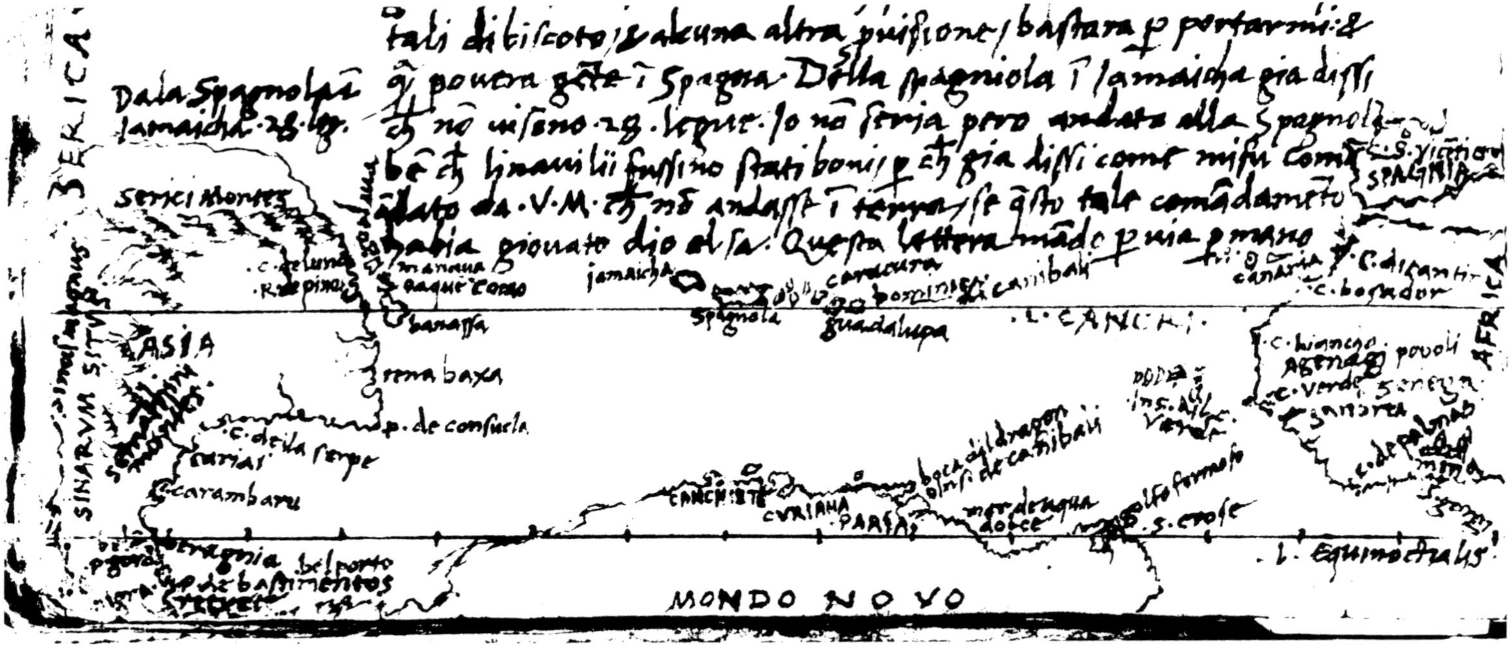

PLATE 21: Alessandro Zorzi's sketch-map, c. 1506. *Though South America is designated as "Mondo Novo," Central America and Southeast Asia are portrayed as one and the same.*

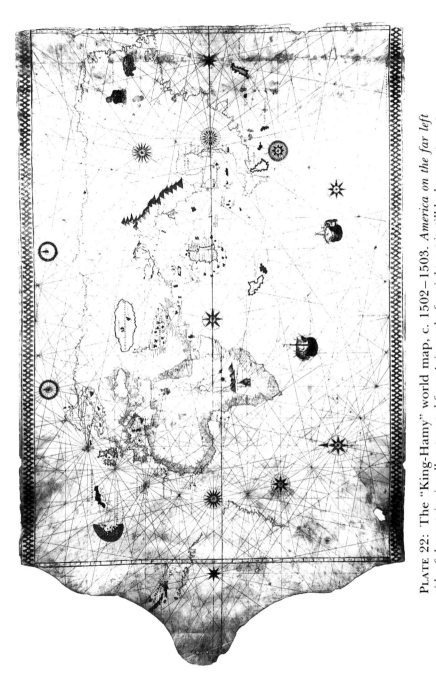

PLATE 22: The "King-Hamy" world map, c. 1502–1503. *America on the far left side of the map is visually separated from Asia on the far right, yet it still lacks a west coast. Notice the tentative delineation of its shorelines.*

Die newe Inſelen ſo zů unſern zeiten durch die künig von Hiſpania im groſſen Oceano gefunden ſinde.

PLATE 23: Sebastian Münster's map of the new islands, 1544. *Though America is portrayed as absolutely distinct from Asia, Japan is still placed just off the shores of Mexico.*

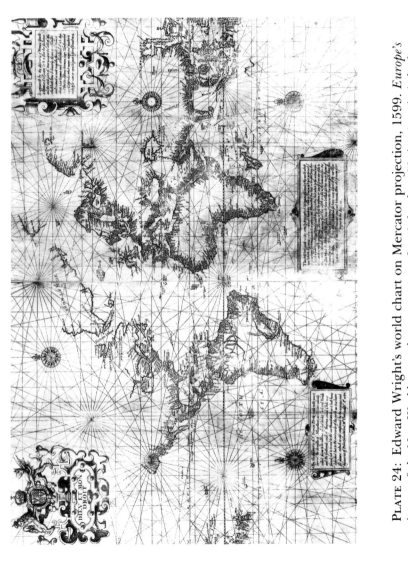

PLATE 24: Edward Wright's world chart on Mercator projection, 1599. *Europe's image of the New World more than a century after Columbus. Notice America's broken shoreline in the northwest, beyond California.*

—55°

—25°

—15°

—10°

5°

MALUCA

—0

—5°

BABAY

JAAVA

SIMVABAA

FOGO

POLYTARIA

—10°

—15°

PLATE 25: Anonymous Portuguese map of the Pacific, c. 1513. *An extremely narrow Pacific Ocean. The Moluccas (Spice Islands) are placed midway between America and Asia.*

PLATE 26: Isaac Tirion's map of the Arctic Pole, c. 1751. *Notice the huge gap still separating Alaska from Oregon.*

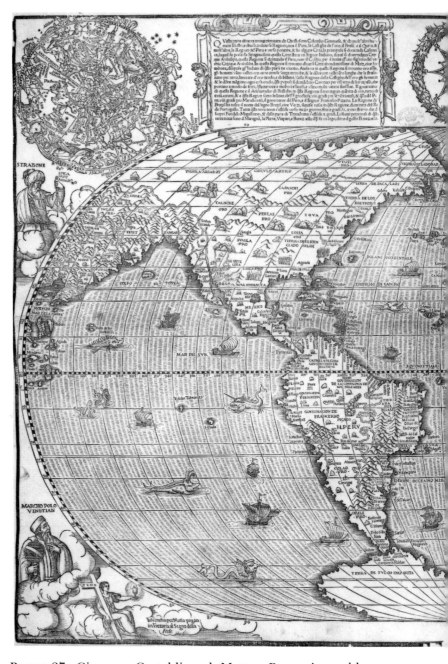

PLATE 27: Giacomo Gastaldi and Matteo Pagano's world map, c. 1550. *North America and Asia are still portrayed as joined by a land bridge. Notice the lions and elephants roaming North America.*

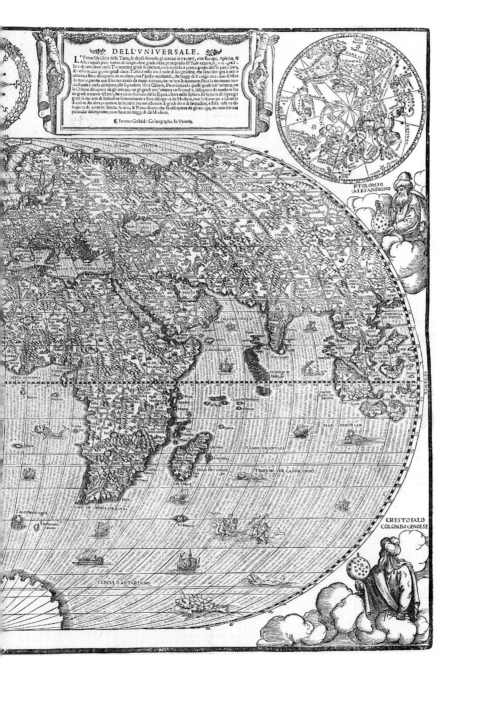

DELL'VNIVERSALE.

L'Vniuersale Orbe della Terra, fu diuisa secondo gli antiqui in tre parti, cioè Europa, Aphrica, &
Asia, i quali puoi hanno di longitudine gradi.dcccc. principiado all'Isole estreme, il, i, i a, i gradi.
Et e di amitudine verso Tra montana gradi.lxiiiiture, cominciando à primo grado dell'e pimo. i, i a, &,
& verso mezo giorno gradi.dxcv. Tutto il resto è ui il vede di loagitudine, che sono dixi gra li centa
ottoana, è stato discoperto da moderni, tra l'Indie occidentali, che hoggi di il vulgo chiama il Mon
do nuouo, perche non è fin mai stato da niuno antiquo, che ne fate le mentione. Pero lo chiamano nuo
uo, sopra le e verso occidente, alle Capradote Isole Canane, Pero stimando quelli gradi cent' ottana, ver
lo Chinate discoperti da gli antiqui, on gli gradi cent' ottaua verso ponete, discoperti da moderni sa
no gradi trecento lestante, che è tanto il ciruolo detto Equinoctiale nella Sphera, Et la parte di sopra gli
gradi le tra arre di latitud ne settentrionale è stata discoperta da Moderni, cioe la Norue gia, è Gentian
di xxo m.lte altre prouince, Et la parte piu meridionale li gradi dxcv di latitudine, è stata anch'ora di
scoperta di moderni, Strabc Arano, & Plino dicono che si è discoperto da gli antiqui, ma non si troua
particular descriptione, come hauemo hoggi di da M.sderni.

¶ Iacomo Gastaldi, Co'mographo. In Venetia.

PLATE 28: Giovanni Vavassore's 1558 copy of Caspar Vopell's 1545 world map (detail). *North America and Northeast Asia are portrayed as one and the same.*

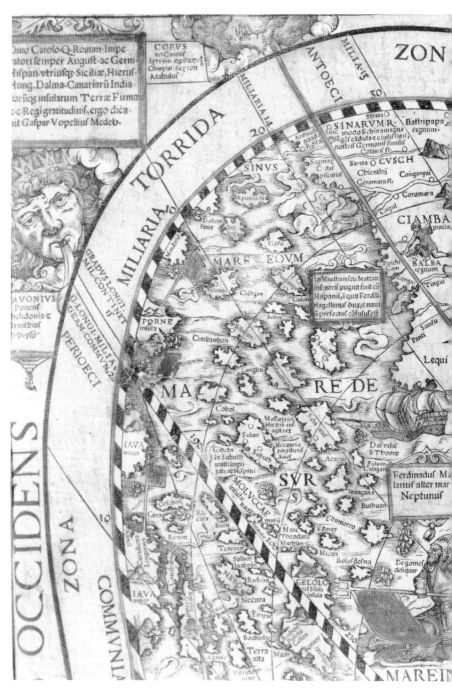

PLATE 29: Giovanni Vavassore's 1558 copy of Caspar Vopell's 1545 world map (detail). *Mexico and China are portrayed as one and the same. Notice how Chinese and Mexican place-names interpenetrate one another.*

EBECCIVS

GRADVS MIL. GERMA. II

GER 40

TEMPERATA

MILI. 10 50

MILIARIA 9 MANICA 8 60

CIRCIVS
In Narbonensi prouincia
Ventorp clarissimus nec
ulli violentia inferior.
θρασκιας αγυμοντης.

A. PR.
latræ Serpen
pro delitijs

PEIM·P·
Incolæ Mabus
metam
Peim met.

PYGMEOR
O regio

CAMVL·P·
Linguam habet pro
priam & manu cultrip

AM·P·
alij
ani

CIARTAM
Reperiuntur hic in flu-
minibus alpides et
Calcedonij

Ciartam
metrop

COTAMP·
Incolæ Mabumetum
colunt, bombicij copia
habentes vinesq;mul
tas.

ERGIMVL·R·
Sunt hic Christiani
Nestoriani Idolatræ
Mabumetani

CHINCHINTA
Incolarum alij Christiani, alij
metam alij Idolatræ, alij
teis q; habentes.

CANICLV·P·
Hic magna margaritarum
aurig copia.

Cotam
metrop.

DIA ORIEN TALIS CERGVT

Hæc regna magno
Cham subiecta.

EZINA Incolæ idolatri
lui et onagris se
bus gaudent.

CARAIA
Hæc septem ha
bet regna

THEBETH
Hæc octo continet
regna

EGRIGAIA
Sunt hic Christiani
Nestoriani

NAVIGVIR
Opulentum et omni
um auro & serico
abundant

Arcladam pro.

TANGVT·PR·
Hic muschum optimum
inuenitur

Omon
lacus

Iacimetrop.

BANGALA
Incolæ habent regem et linguã
propriam & idolatræ, abundant spico, galanga, zinzibere
& saccaro

Sigui
THOLOMA
Incolæ linguam habent
propriam, aurig copia
& ciuitates multas.

Lop

LOP
Incolæ Mabum
tum colunt.

CVNCHIP
Hic serici, leonum et
ursorumq; copia.

Singui
empori

Cangigu

AMV·P·
Tam virig feminæ
monilia gestant aurea

Hic auri & aroma
tum copia

TEVDVCH
Sunt hic Christiani qui
primat tenent, tribu
ta pendentes magno Cham.

GINGVI PR
Incolæ Idola colunt, serici
habet, faciunt q; pãnos
tes de arborum cortici

ASIA MAGNA

QVEMQVINAFV
Hoc regnum fuit olim opulentum et
maximum, abundat sericis.

In hac Asia magna Dei Euã
gelium indies et indies indicis
Hispanos seminantur.

CAMBALV·R·
Cambalu
regalis

CACAFU

SINDIFV
regnum regem
ditissimum ac
potentem ha
bens.

Sindinfu

Alchalech
mangi prouincia

Atcin

Cathay regnum maximu
omnium mundi toti
us repletum genti
bus, diuitijsq;
infinitis

TAMACHO
Olim Tanguth dicta, faciut
hic pãnos præciosos ad et
findones.
Portum de Panuco
Hispani obtinent

CHATAY·R·

MESSIGO
Temporibus uno
finmans vo
cabula annotauit

FVGVI·O
Cresset hic reu
barbarum incopia
argingui

Cangli

Tonalo
P. d. Panuco

MANGI
prouincia
nouem continet
regna

Cingianfu

Cunao
AMCEL
vel. S. Michaelis
prouincia

MASAMACO

R. daobiæ·P·

Vnquen

Quel insu

Singui

Tesqua

ACVLVA CAN
prouincia

Singui

Archi
dena

Maimalte
beque

HISPANIA
Metico
prouincia

TEMIXTITAN
Regnum e ciuitas

CALCO
Gratio

CVLVA

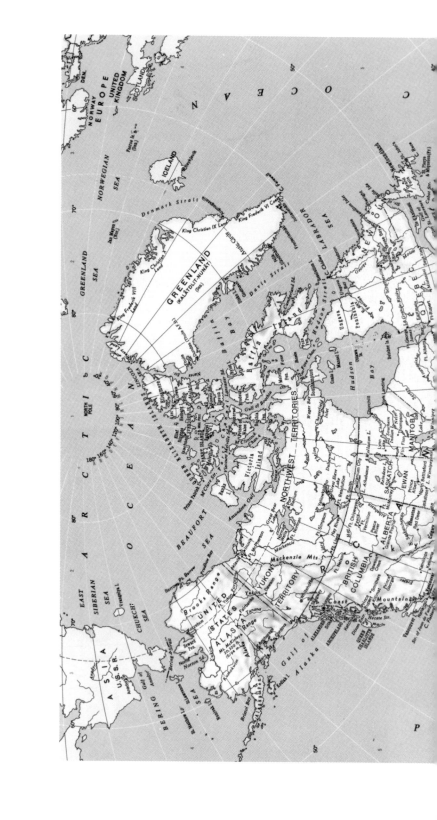

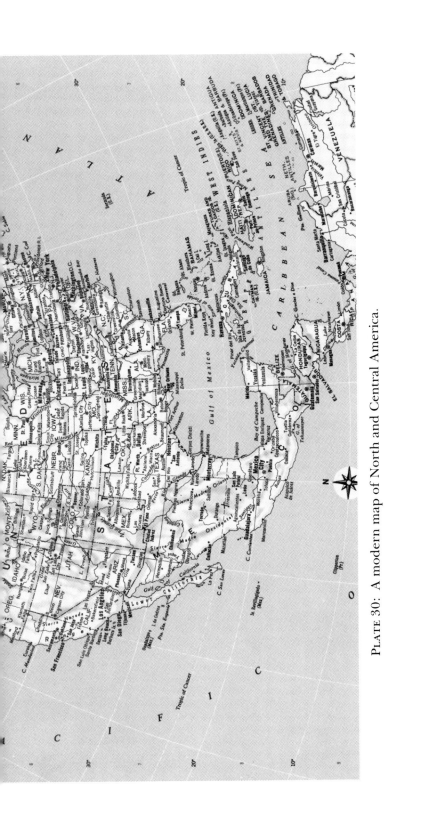

PLATE 30: A modern map of North and Central America.

PLATE 31: The "Kunstmann IV" map of the Western Hemisphere, c. 1518. *North and Central America's shorelines are still tentatively broken.*

pucci, Duarte Pacheco Pereira, and Martin Waldseemüller apart from their contemporaries. These intellectual pioneers deserve a special place in the annals of the discovery of America for having had the intellectual audacity to speculate, long before it was empirically demonstrated, that the newly discovered land beyond the Atlantic was in fact something quite distinct and separate from Asia. Instead of viewing it, as Columbus did, as a mere extension of the familiar (that is, as a group of islands off the shores of China), they went boldly ahead and redefined it as something entirely new! Most remarkable, in this regard, was Pacheco's and Waldseemüller's explicit identification of America as a fourth continent. Given the traditional image of a tricontinental world, it marked a complete shift in Europe's entire cosmographic perception.

It certainly took a lot of intellectual guts literally to rediscover "the Indies" as a New World that was absolutely distinct and separate from the Old. When we come to examine Vespucci's, Pacheco's, and Waldseemüller's accomplishments, we must not overlook the fact that the way we now envision America presupposes a well-established modern cosmography that these intellectual pioneers practically had to invent themselves. And inventing it entailed having fully to abandon its classical precursor. It is hard not to admire the intellectual audacity of Vespucci, Pacheco, and Waldseemüller, their readiness to stick to their convictions even when it meant having to give up the security offered by convention, to forgo a well-established system of interpretation when an alternative one had yet to be invented.

As Waldseemüller himself warned his readers:

> In designing the sheets of our world-map we have not followed Ptolemy in every respect, particularly as regards the new lands. . . . [T]hose who notice this ought not to find fault with us, for we have

done so purposely. . . . Ptolemy himself, in the fifth
chapter of his first book, says that he was not ac-
quainted with all parts of the continent. . . . It has
been necessary therefore . . . to pay more attention
to the information gathered in our own times.[41]

Waldseemüller's willingness to transgress the secure confines of tradition is even more remarkable when contrasted
with the stubborn efforts of many of his contemporaries to
force the new discoveries into the old cosmography. It certainly requires a great deal of intellectual courage to recognize that a well-established familiar paradigm can no
longer account for some new findings and to develop instead an alternative system of interpretation that might.

Their readiness to relinquish the structures offered by
classical cosmography was a clear testimony to Vespucci's,
Pacheco's, and Waldseemüller's creativity. Creativity, which
almost by definition precludes the acceptance of any structure as a given, presupposes enough flexibility to forgo familiar mental structures on the realization that the reality
one experiences may warrant new ones.[42] In their effort to
break away from the mental confines of medieval dogma,
these intellectual pioneers clearly embodied the distinctive
spirit of the Renaissance.

Denial

Vespucci's, Pacheco's, and Waldseemüller's formidable display of intellectual guts is even more remarkable when
contrasted with the much-more-common response of their

contemporaries to the totally unexpected discovery of a previously unknown continent beyond the Atlantic—namely denial. The antithesis of intellectual courage, denial is a way of resisting the unfamiliar by forcing it into familiar mental niches, thereby practically denying its novelty or unusualness.

New facts that call for a total revision of the way we usually organize the world in our mind obviously create difficult cognitive problems. In order to avoid those and keep the established mental structures intact, we often try to deny the novelty or unusualness of such inconvenient facts. Consider, for example, the way people often force the sexually or racially ambiguous into one of the conventional categories available[43] or the way scientists typically respond to anomalies that defy their established paradigms of explaining reality.[44]

The practical annihilation of classical cosmography by ideas such as Vespucci's and Waldseemüller's clearly threatened many Europeans, as it left them without the security offered by familiar structures. The best way to avoid that, of course, was to deny the idea that the newly discovered lands beyond the Atlantic could possibly be anything other than Asia. And indeed, long after 1492 many Europeans still kept insisting that the New World was either totally identical with, or at least somehow attached to, the Orient.

Thus, for example, when Vicente Yáñez Pinzón first landed in Brazil in 1500, he thought he was "beyonde the citie of Cathay and the coastes of Easte India beyonde the ryver of Ganges."[45] In fact, on his first voyage to South America a year earlier, even Vespucci himself kept searching in the Amazon Delta for Catigara (Cattigara), the southeasternmost point of the Asian mainland on Ptolemaic maps.[46]

Yet the most glaring example of such a response to the

discovery of the New World was that of Christopher Columbus. His cosmographic ideas about America capture
the spirit of denial in its purest form.

◆

The most distinctive characteristic of the way Columbus
interpreted his own discoveries beyond the Atlantic was
the stubborn manner in which he tried to force practically everything he encountered on his first voyage to
America into the traditional image of a tricontinental
world. Throughout that voyage, it was his rigid preconceptions, evidently unaffected by the facts themselves, that
dictated his cosmographic interpretation of what he actually found beyond the "Ocean Sea." "Columbus was,
indeed, under the domination of a fixed idea, and ratio-
nalized all his experiences into harmony with his earnest
wishes."[47]

Most remarkable in this regard were Columbus's relent-
less efforts to force the totally unfamiliar new continent
into the familiar contours of the Old World. As evident
from the entry in his diary on the day of his very first encounter with it, Columbus identified America right from
the start as "the Indies" and its inhabitants as "Indians."[48]
Cuba, Hispaniola, and the Bahamas, whose existence had
until then been virtually unknown in Europe, thus became
in his mind rather familiar entities. They were among the
7,440 islands lying, according to Marco Polo, in the China
Sea off the shores of Asia.[49] His strong belief that he had
actually reached the Orient is also quite evident from the
repeated references in the diary to Japan (Cipango) as well
as to the Great Khan.[50]

Throughout his first voyage to America, Columbus's
wishful thinking clearly made him see and hear only what
he wanted to. Thus, for example, he would quite casually

co-opt native words and names into his own system of expectations, being quite certain that his informants were indeed telling him about the Great Khan ("whom they call *cavila*")[51] or Japan ("which they call Cybao").[52] Such self-delusion also led him to "find" in the New World distinctively Old World plants such as aloe, nutmeg, cinnamon, and rhubarb.[53]

Unfortunately for Columbus, nor did his second voyage to America help to confirm its Asian identity. In fact, the more he discovered there the less he could fit it into Europe's image of the Orient. The new lands he encountered (Guadeloupe, the Virgin Islands, Puerto Rico, Jamaica) were all islands, and the elusive Asian mainland was still nowhere in sight.

Yet the conviction that he had in fact reached Asia had clearly become an *idée fixe* for Columbus and, further utilizing his tremendous power of denial, he continued to do everything he could to sustain it. Nowhere was that more blatantly evident than in the way he dealt with Cuba.

On his first voyage to the Caribbean, Columbus had already heard from the natives that Cuba was only an island, yet he stubbornly kept insisting that it was part of the Asian mainland and that he was in fact only one hundred leagues away from the Chinese cities of Zayto and Quinsay.[54] In 1494, on his second voyage there, he returned to the island and, still contending that it was the Asian mainland and that he was not far from the Golden Chersonese (the Malay Peninsula),[55] tried to follow its southern coastline westward until he would reach the Chinese province of Cathay.[56] Taking the tip of the island (Cape Maisí) to be the easternmost extension of the Old World,[57] he named it "Alpha and Omega" to express his conviction that it was the end of the West as well as the beginning of the East— the point where the two hemispheres actually met![58]

As we all know, Columbus never reached China on that voyage (or indeed on any other). However, since he stopped and turned back before reaching the end of the island, he could still claim that it was part of the Asian mainland. In order to do so, however, he first had to force all his men to sign a statement under oath (on the punishment of having their tongues cut out if they ever broke it) to the effect that it was indeed part of the mainland and that there was no need to sail farther west to prove it.[59]

On August 5, 1498, on his third voyage to America, Columbus finally reached what had so stubbornly eluded him on his two earlier voyages across the Atlantic: For the first time since he had begun exploring the New World six years earlier, he managed to discover not just another island but the long-anticipated mainland (*tierra firme*) itself.

It was the enormous size of the Orinoco Delta that convinced Columbus that the Paria Peninsula in what is now Venezuela was indeed part of the mainland and not just another island. As his great chronicler Bartolomé de Las Casas recounts it, on August 13 Columbus

> decided to go westward along the mainland coast, still believing it was the island of Gracia . . . and see whence came such a great flood of water, and if it proceeded from rivers as the sailors asserted, the which he says he did not believe, for he knew that neither the Ganges nor the Euphrates nor the Nile carried so much fresh water. The consideration that moved him was that he did not see lands large enough to provide a source for such large rivers, *unless, he says, this land is a continent.* These are his words.[60]

And indeed, the following day Columbus already noted in his journal: "I have come to believe that this is *a mighty continent that was hitherto unknown.* I am strongly supported in this view by the reason of this great river."[61]

The statement that it was "hitherto unknown" suggests that Columbus must have realized right from the start that "the land of Paria" was not part of the Asian mainland. His usual strategy of forcing the totally unfamiliar new lands he discovered beyond the Atlantic into the more familiar image of the Orient was evidently no longer a viable option, given his obvious inability to fit the great northeastward-flowing Orinoco River into that image. And yet, utilizing once again his seemingly inexhaustible power of denial, Columbus soon found an original solution that allowed him to "place" this new continent in a way that would still not compel him to transgress the confines of the traditional image of the tricontinental Old World.

The solution was to identify South America as the "Earthly Paradise," which, according to the Bible as well as many Christian theologians, one might indeed expect to find in the Orient.[62] On August 17 he first notes in his journal that he might have in fact reached the "Terrestrial Paradise," which "all men say" is "at the end of the Orient."[63] In the letter he sent to Ferdinand and Isabella from Hispaniola two months later, the dramatic discovery of this paradise "at the end of the East" was already presented as an established fact.[64]

His fourth and final voyage to America was supposed to provide Columbus with a clearer picture of the yet unexplored area between the "Earthly Paradise" and Cuba, which he still maintained was part of the Asian mainland. And when in August 1502 he finally encountered for the first time the central part of the American mainland (in northern Honduras), it again seemed to him that he must have reached the Orient. In fact, as he sailed around Cabo Gracias a Dios and began to proceed southward along the Nicaraguan coast, he was positive that he was in Southeast Asia and was actually searching for the strait that supposedly separated Sumatra from the Malay Peninsula—a strait that, following Marco Polo, would definitely lead him finally into the Indian Ocean.[65]

As we all know, Columbus never found the elusive strait leading from the Caribbean (which he still believed was the China Sea) into the Indian Ocean—a strait that of course does not exist. Moreover, by the time he reached Costa Rica, the general trend of the coast started to change from southward to southeastward. And when in November he finally reached the port of Retrete in Panama, he found evidence that Rodrigo de Bastidas, coming from the opposite direction (that is, from Colombia), had gotten there several months before him.[66] Having been at Hispaniola earlier that summer just when Bastidas stopped there on his way back to Spain,[67] Columbus could have known by then that a continuous coast in fact joins Panama and his "Earthly Paradise" on the northern shore of South America (which, in the wake of the voyages of Vespucci, Pinzón, Diego de Lepe, and Vélez de Mendoza between 1499 and 1501, was already known by that time to extend as far as the eastern coast of Brazil). In other words, he could very well have realized at that point that the central and southern parts of America were indeed parts of a single continuous landmass.

That finding would most probably have led a more intellectually courageous cosmographer seriously to reconsider his original image of Central America and conclude that perhaps it was not Southeast Asia after all. Yet Columbus, demonstrating once again his tremendous power of denial, chose instead to try to force the new findings into his old system of cosmographic preconceptions rather than use them to update it. In the letter he wrote from Jamaica in July 1503 to Ferdinand and Isabella describing his encounter with Central America, he claimed that the province of Veragua in Panama was in fact the biblical land of Ophir, from which Solomon had twenty-five centuries earlier brought to Jerusalem the "gold of the Indies" to help build the First Temple![68] (Ten years earlier, in fact, he had already suggested that Cuba and Ophir were one and the same.[69]) His wishful thinking likewise led him to "hear" the natives of Costa Rica telling him about the nearby Southeast Asian province of Ciamba as well as about the great Ganges River, which, supposedly according to their own reports, was only ten days' journey from the Central American province of Ciguare.[70]

To enhance the cosmographic credibility of his claim that Central America and Southeast Asia were one and the same, Columbus also invoked in his letter to the Spanish monarchs the authority of Marinus of Tyre, a second-century geographer whose estimate of the distance separating western Europe from eastern Asia was forty-five degrees shorter than Ptolemy's.[71] (In favoring Marinus's estimate over that of his famous contemporary, Columbus was actually reiterating an opinion voiced some ninety years earlier by the French theologian Cardinal Pierre d'Ailly in *Imago Mundi*, the most extensively annotated book he owned.[72]) "The world is small," he thus announced to his royal sponsors, with only one seventh of its surface actually covered by water.[73] That, of course, would help explain how he could have reached the Orient in such a short time. (The very same year, incidentally, the canon of Seville, Rodrigo de Santaella, correctly placed the West Indies on the opposite side of the globe from India.[74])

Trying to breathe new life into a fourteen-hundred-year-old theory instead of honestly examining the massive evidence lying all around him was not an unusual course of action for Christopher Columbus. It certainly fitted his usual tendency to try to force reality into his own preconceptions. Such a tendency, in fact, underlay some of the major "negative decisions" he made in his career as an explorer.

Consider, for example, his decision to abort his westward advance along the southern shore of Cuba in June 1494. Luckily for Columbus, he made that historic decision when he was only fifty miles away from the western end of the island. Had he gone on, he would have inevitably found out to his great disappointment that it was indeed an island and not part of the Asian mainland as he so much wanted to believe. His decision to turn back, of course, may indeed have been precipitated by the grumbling of his men as well as the fact that his caravels were leaky and his provisions low. Nevertheless, though he still spent several more years in the Caribbean between 1494 and 1504, never again did he ever attempt fully to resolve the mystery of Cuba. Doing that would not have presented any great technical challenge for Columbus, but it would certainly have resulted in a profound psychological shock for him. And so, instead of further exploring Cuba and eventually finding out that it was in fact only an island, he rather conveniently satisfied his curiosity by forcing his crew to sign a statement to the effect that it was part of the Asian mainland and stubbornly held on to his *ideé fixe*.

Such intellectual timidity in the face of new facts also explains another major "negative decision" made by Columbus on his third voyage to the New World. After all, when he discovered South America in 1498, he could have proceeded farther west- or eastward in an honest effort to determine its true cosmographic nature and identity, as Ojeda, Vespucci, and others in fact soon did. Instead, he was satisfied merely to identify it as the "Earthly Paradise" and headed straight to the more familiar "Indies." Nor did he ever try to return to South America thereafter.

In short, it was not his poor health, lack of provisions, inhospitable weather, leaky caravels, mutinous crews, and hostile Caribs alone that kept Christopher Columbus from realizing the full cosmographic significance of his historic encounters with the West Indies, South America, and Central America. Nor was it just sheer bad luck or the mere lack of effective communication with the indigenous population of Panama (as some of his more sympathetic biographers have suggested[75]) that prevented him from reaching the Pacific eleven years before Balboa. It was also his intellectual timidity, his obvious fear of generating any new evidence that might challenge his theories. The man who has traditionally been portrayed as the quintessential symbol of venturesomeness and audacity was in fact highly intimidated by the very prospect of discovering new facts that might upset his stubbornly held convictions.

Each of Columbus's four voyages to America, of course, had its own set of specific circumstances that may help explain the particular decisions he made as well as his failure fully to understand what he had actually accomplished on them. Examining all of them together, however, reveals a clear pattern of avoidance and denial in the face of inconvenient evidence that was not easily reconcilable with his expectations. That pattern becomes even more striking when we contrast the actual geographical scope of Columbus's voyages with that of Vespucci's, Magellan's, or Cook's. In fact, for nearly twelve years, he literally kept going around the Caribbean in circles in an effort not to transgress the confines of his highly restrictive cosmographic vision. As a result, though he certainly played a critical role in the history of America's physical discovery by Europe, he played only a negative one in the history of its mental discovery.

◆

The most explicit visual expression of Columbus's own image of America after his fourth and final voyage are three sketch-maps made in 1506 by Alessandro Zorzi to embellish an Italian translation of Columbus's 1503 letter from Jamaica (the *Lettera Rarissima*) to Ferdinand and Isabella, which his brother Bartholomew brought to Rome to present to Pope Julius II. Though ultimately drawn by Zorzi, possibly with Bartholomew's help, the maps were most probably based on drawings made by Columbus himself.[76] As evident from the continuous coastline joining Brazil and Honduras on one of them, Columbus must have realized before he died that Central and South America were indeed parts of a single continent. Yet the map also shows that he did not envision that continent as distinct from Asia, since it quite clearly features Central America and Southeast Asia as one and the same (Plate 21). The very same landmass identified on the left side of the map as "Asia" and bearing unmistakably Oriental Ptolemaic place-names such as Cattigra (Catigara) and Serica is nevertheless also dotted with names mentioned in Columbus's letter in connection with places he visited in Costa Rica and Panama—Cariai, Carambaru, Belporto, Bastimentos, Beragnia (Veragua), and Retrete. According to Columbus, China and Brazil were clearly parts of a single continent.

The cosmographic vision of America as part of Asia did not die with Columbus. Three well-known maps produced in Italy soon after his death—Contarini's (1506) and Ruysch's (1507) world maps and Rosselli's 1508 marine chart of the world (Plates 14, 2, and 13)—capture quite vividly the reluctance of many of his contemporaries to accept the fact that the new continent was indeed quite distinct and separate from Asia. (The first two were also the earliest *printed* maps to show America and, as such, had an unprecedented circulation and probably played a major role in the way many Europeans initially came to envision it. Ruysch's map also had considerable "official" authority, having been included in the first post-1492 editions of Ptolemy's *Geography*.) They offer us a glaring visual expression of the conservative effort of Europe to deny the "novelty" of the New World.

As we have seen, the early Portuguese world maps from 1502–1505 (Cantino, Kunstmann II, King-Hamy, Caveri, Oliveriana [Pesaro]) helped promote the separateness of the New World from the Orient in Europe's mind by featuring America on their left side and eastern Asia on their right. By contrast, following the cartographic precedent set by Zorzi in his 1506 sketch-map, Contarini, Ruysch, and Rosselli all placed eastern Asia on the left side of their world maps, thereby allowing for a visual blending of East and West.

Thus essentially denying the "novelty" of the New World, all three explicitly portrayed the newly discovered lands beyond the Atlantic as part of Asia. Like Zorzi, Rosselli basically placed Central America in Southeast Asia (*west* of Indonesia), filling present-day Vietnam with Honduran, Costa Rican, and Panamanian place-names taken from Columbus's 1503 letter—Cabo Gracias a Dios, Bastimentos, Belpuerto, and Retrete (Plate 13)! Contarini, too, noted on one of the legends on his map that on his voyage to Central America, Columbus had in fact reached the southern Chinese province of Ciamba.

The three also portrayed the two great oceans lying west of Europe and east of Asia as practically one and the same. Accordingly, they put Japan in the Caribbean, along

with the other "Indies" discovered by Columbus. Contarini thus placed "Zippangu" right next to Cuba (Plate 14)
while Ruysch went even further, claiming on one of the
legends on his map that Marco Polo's Cipango was in fact
Hispaniola.

The total blending of America and Asia becomes even
more pronounced as we proceed northward. As we shall
see later, in cutting northeastern Asia off at the edge of
their maps, both the producer of the Cantino map and
Caveri basically still left open the possibility that it might
in fact be connected to North America (Plates 3 and 8).
Contarini, Ruysch, and Rosselli made that possibility a reality by explicitly placing the recent Portuguese discoveries
in Canada on the northeastern tip of Tartary (as did Vesconte de Maggiolo in his 1511 world map), thus literally
portraying North America as an Asian promontory (Plates
14, 2, and 13).

Given Europe's rather sketchy picture of both northeastern Asia and northwestern America until the eighteenth
century, it took a long time before it could be absolutely
certain that the two continents were indeed not connected
to each other in the north. In fact, even Balboa's and Magellan's discoveries in the Pacific did not fully destroy the
conservative image of America as an extension of Asia.
While they clearly demonstrated that the New World and
the Orient were absolutely separate from each other in the
south, they still left open the possibility that they might
somehow be connected in the yet unexplored north. And
though there was no evidence to support such a conten-
tion, neither was there any that could altogether eliminate
it once and for all.

Cortés's first encounter with the Aztec Empire in 1519

certainly heightened the suspicion that China and Mexico were indeed one and the same, leading some cosmographers, for example, to equate the Yucatán Peninsula with Mangi or Japan.[77] Such a fanciful blending of East and West, first expressed visually on a small map of the Western Hemisphere made in 1527 by Franciscus Monachus, is quite glaringly featured on the c. 1528 Gilt globe (Plates 19 and 20), Oronce Finé's 1534 cordiform world map (Plate 1), the c. 1535 Nuremberg globe gores (Plate 12), and Giovanni Vavassore's 1558 copy of Caspar Vopell's 1545 world map (Plates 28 and 29), all of which portray Mexico and China as essentially one and the same. These and similar maps, globes, and globe gores—the c. 1530 Nancy globe, Finé's 1531 world map, an anonymous map from c. 1535, Vopell's 1536 globe, Haggi Ahmed's 1559 world map, Giovanni Cimerlino's 1566 world map, Bernard van den Putte's 1570 copy of Vopell's 1545 map, Georg Braun's 1574 world map, Mario Cartaro's 1579 astronomical diagrams, Giacomo Franco's 1587 world map—all feature New Spain (*Hispania Nova*) right next to the provinces of Mangi and Cathay, practically interspersing distinctively Chinese place-names such as Marco Polo's Zaiton and Quinsay with unmistakably Mexican ones like Messigo, Temixtitan, Aculuacan, and Teniscumatan (Plates 19, 20, 1, 12, and 29). As Orient and Occident literally intermingle, we also see Chinese rivers flowing into the Gulf of Mexico, occasionally identified in fact as the China Sea (*Mare Cathayum*).[78]

The interpenetration of Asia and America on these maps and globes is by no means confined to Mexico alone. Some of them, for example, explicitly equate Hispaniola or the Yucatán with Japan and place Southeast Asian islands like Java, Timor, and the Moluccas just off the west coast of Central America (Plates 20, 1, 12, and 29).[79] They likewise place the Southeast Asian province of Ciamba in Nicaragua[80] and Ptolemy's Catigara (*Cattigora*) in Ecuador (Plates 20, 1, and 12).[81] By the same token, in the north they place *Asia Orientalis* right next to French Canada (*Terra Francesca*) (Plate 28).

As late as the 1570s and even 1580s, we still find cartographic representations of America that are clearly modeled after Finé's and Vopell's originals.[82] Yet by the 1550s a somewhat different cosmographic vision of the New World starts to dominate the European mapscape: it now no longer fully blends with the Orient yet is still attached to it by a northern land bridge.

By the late 1530s probably only a very few Europeans in Mexico still believed it was China. In 1539, however, a French Franciscan friar, Marcos de Niza, returned to Mexico from the upper Rio Grande with dazzling stories about camels, elephants, and people with silk clothing,[83] and the following year a Spanish expedition led by Francisco Vásquez de Coronado followed those rumors into Arizona, New Mexico, and even Kansas. As one of Coronado's men reported on their return, "New Spain is part of a continuous continent with Peru, as well as with greater India or China. There is no strait in this region to divide it."[84] He also added that the Pueblo Indians must have come to the Southwest from "Greater India" simply by crossing the Rocky Mountains.[85] In the same vein, basically denying the unusualness of the unfamiliar in a style remarkably evocative of Columbus, a Franciscan friar who participated in Coronado's expedition identified the first buffalo ever seen by Europeans as the same kind of cattle Marco Polo had reported seeing on his travels in the Orient![86]

It was evidently such reports that led Venetian cosmographer Giacomo Gastaldi to introduce on his 1546 world map a northern land bridge joining Asia and America. In sharp contrast to Vopell's world map from the year before (Plate 29), China no longer blends on that map with Mexico and the actual contact between East and West takes place only in the yet unexplored regions north of California and the Southwest. Yet those regions are definitely an extension of Asia, and in Gastaldi and Matteo Pagano's c. 1550 world map one already sees lions and elephants roaming the United States and Canada (Plate 27)!

This new vision of the New World appeared in several other Gastaldi maps in the 1540s and 1550s (for example, his 1548 and 1555 world maps) and was soon adopted by other European cartographers as well—Giorgio Calapoda in 1555; Paolo Forlani in 1560, 1562, 1565, and 1570; Girolamo Ruscelli in 1561; Benito Arias in 1571; and Tommaso Porcacchi in 1572. Some of the maps featuring it (Gastaldi's 1548 and Ruscelli's 1561 world maps, for example) were even included in new editions of Ptolemy's *Geography*. A few of those maps—Joannes Myritius's 1590 world map and the 1594 and 1599 reissues of Forlani's 1560 and Ruscelli's 1561 world maps,[87] for example—were in fact printed as late as the 1590s. In other words, more than a full century after its "discovery" by Columbus, the New World was still envisioned by many Europeans as at least partly attached to the Old. Old ideas certainly die hard.

Ambivalence

Forlani's and Ruscelli's world maps represent the final death throes of the conservative attempt of Europe to

deny the "novelty" of the New World and its absolute separateness from the Old, since they are a glaring expression
of the stubborn wish to maintain at least some semblance
of connectedness between them. After all, the America
they portray is technically still part of the Old World. Indeed, the fictitious land bridge they feature joining Asia
and North America was the very last piece of string on
which the entire classical tricontinental cosmography was
still hanging.

The very essence of a New World, of course, lies in its
being perceived as absolutely distinct and separate from
the Old. The process of establishing the separateness of
any entity from other entities, however, is very often offset
by the diametrically opposite wish to preserve at least some
connectedness between them. In other words, encounter-
ing a "new" entity often also entails some ambivalence
about its novelty.

A perfect case in point is the way we normally establish
our distinctive identity as individuals. The process of individuation, of course, is also one of separation, since
forming our own identity inevitably presupposes a readiness to separate ourselves psychologically from our parents and let go of our primal illusion that we are literally
attached to them.[88] Yet the need to distance ourselves from
our parents in order to establish an identity separate from
theirs is usually offset by an equally strong need still to feel
attached to something. It is precisely the fact that they provide us with the feeling that we are not entirely alone in
the world that makes love and friendship so important to
us especially during adolescence, when we try to break
away from our parents. Our insatiable lifelong need for
intimacy is a clear expression of the great yearning for
connectedness that accompanies the neverending process
of individuation.[89]

The stubborn effort to deny the separateness of the New World from the Old—visually expressed first in the attempt to portray them on maps and globes as virtually blending into and later as at least partly attached to each other—is highly evocative of our desperate need to cling to something so as not to feel totally lost in the world as we make our first attempts to establish a separate identity. The mythical land bridge that keeps the young, new continent literally attached to the Old World on maps such as Gastaldi's is the functional equivalent of the security blankets and other "transitional objects"[90] that help us feel connected to something at the very same time that we try to establish some psychological distance from our parents. The discovery of the Bering Strait was thus the final rupture of the symbolic umbilical cord that still connected the New World to the Old in Europe's fantasies.

The mythical land bridge joining Asia and North America was only one expression of sixteenth-century Europe's general ambivalence about the cosmographic status of the newly discovered lands beyond the Atlantic. Such ambivalence need not surprise us; after all, serious doubts very often haunt both avant-gardists and conservatives, and our actual response to what is novel usually lies somewhere between the two extremes of innovation and denial.

Given the rather sketchy picture of America that was actually available to them throughout the sixteenth century, it was only natural that many Europeans would indeed feel somewhat ambivalent about its actual cosmographic status relative to Asia. And not even those who were ready to examine honestly the new evidence coming from across the Atlantic, rather than simply deny it, were necessarily always ready to give up totally their traditional image of a tricontinental world right away. As a result they were somewhat equivocal about the "novelty" of the New World.

Many of the maps made during the sixteenth century thus embody the various efforts by Europe to reconcile the extremes of innovation and denial in its overall response to the understandably traumatic discovery of America.

◆

Among the most glaring expressions of ambivalence, of course, are the double messages we send when we "speak out of both sides of our mouth." However, such mixed feelings may also be displayed cartographically when two conflicting visual statements are made on one and the same map. Thus, for example, on his 1507 world map, Waldseemüller expresses his ambivalence about whether North and South America are indeed parts of a single continent by actually presenting both possibilities. On the map itself he portrays them as separated from each other by a strait. At the same time, on the small inset map above it, he also shows them joined together by a narrow land bridge (Plate 6). Contarini, Ruysch, and Rosselli likewise offer us a curious display of ambivalence by making—each on his own map—two rather conflicting statements about the relations between Asia and America. Though they clearly deny the "novelty" of the New World by presenting North America as part of Asia, they nonetheless also affirm it by portraying South America as a totally separate continent (Plates 14, 2, and 13).

Consider also the narrow strait that separates North America from Asia on most late-sixteenth- and early-seventeenth-century maps and globes. A prophetic anticipation of Bering's historic discovery 167 years later, this purely mythical strait made its cartographic debut on a 1561 world map by Gastaldi, who named it the Strait of Anian. Fellow Venetian cartographers Bolognino Zaltieri (1566) and Giovanni Francesco Camocio (1567) soon incorporated it into their own maps, and it was not long before it appeared even on Mercator's famous 1569 world map. By the late sixteenth century, Gastaldi's fabled Strait of Anian was a common feature on almost every major European globe and world map, and it remained extremely popular throughout the early seventeenth century. (In fact, it was featured as late as 1772 on Didier Robert de Vaugondy's map of North America, which was also included in the 1780 edition of Denis Diderot's *Encyclopédie*.[91])

Since the actual geography of the North Pacific was still virtually unknown at the time these maps and globes were made, their very depictions of this purely fictional strait tell us quite a lot about Europe's deep fantasies about America during the late sixteenth and early seventeenth centuries. At the same time that its appearance seems to express Europe's basic acceptance of America's separateness from Asia, its extreme narrowness also expresses its fundamental ambivalence about this separateness. On the maps and globes featuring the mythical Strait of Anian, the Old World and the New World are essentially portrayed as both detached from, and attached to each other—a perfect visual expression of Europeans' deep cosmographic ambivalence about their actual status relative to each other!

The same ambivalence is also evident from the way Japan is portrayed on early sixteenth-century European maps and globes. Prior to the arrival of the Portuguese in 1542, this island was known to Europe only through secondhand rumors reported by Marco Polo from China two and a half centuries earlier, so its actual distance from America was still virtually unknown. The way in which the two are nevertheless situated relative to each other on early-sixteenth-century maps and globes therefore tells us a lot about Europe's early fantasies about America.

We have seen earlier how conservative European cartographers who basically denied the novelty of the New World often placed Japan ("Zippangu," "Zipangri") somewhere in the Caribbean, along with the other "Indies." Thus, for example, it appears right next to Cuba on the Contarini (1506) world map (Plate 14) and is virtually equated with Hispaniola on the Ruysch (1507) and Finé (1534) maps (Plates 2 and 1) and with the Yucatán on the Gilt (c. 1528) globe and Nuremberg (c. 1535) globe gores (Plates 20 and 12). Yet avant-garde cartographers also evidently assimilated Japan into their images of the New World. In fact, even Waldseemüller, the first man who unequivocally featured America as fully detached from the Orient, nevertheless described Japan in his book as an island "in the *Western* Ocean"[92] and portrayed it on the small inset map above his famous 1507 world map as somewhat closer to the New World than to the Old (Plate 6). (On the large map [Plate 5] it is portrayed on the right side of the map, yet only ten degrees west of Central America.) By the same token, Japan almost touches South America on the c. 1510 Lenox globe and is situated very close to Central America on Glareanus's pre-1520 maps of the Pacific (Plate 11) and the Northern Hemisphere as well as on Stobnicza's 1512 world map, the Schöner (1515 and 1520 [Plate 9]) and Paris (c. 1515) globes, and the Boulengier (c. 1514) and Liechtenstein (c. 1518) globe gores. In fact, even Magellan's voyage across the Pacific in 1520–1521 did not immediately dispel such cosmographic visions. Thus, for example, in Pietro Coppo's and Simon Grynaeus's 1528 and 1532 maps, Japan still appears on the far left side of the map as the westernmost extension of the New World, and even Sebastian Münster (Plate 23) in 1544 and François Demongenet around 1560 still place it just off the shores of Mexico.

Ever since they first heard about it from Marco Polo,

Europeans had always perceived Japan as an extension of the Orient. That is quite evident, for example, from Henricus Martellus's c. 1489 world map, which portrays it on its far right side (Plate 4), as well as from Martin Behaim's 1492 globe, one of whose legends describes it as the richest island "in the east."[93] Its cartographic assimilation into the West in the early sixteenth century is therefore quite revealing, underscoring yet again Europe's fundamental ambivalence about the absolute separateness of the New World from the Orient (even after Magellan's three-month voyage across the Pacific, which clearly offered a rather definitive demonstration of its full extent).

The fact that even people who were obviously committed to a nontraditionalistic cosmography nevertheless felt a need to assimilate Japan into their image of the New World demonstrates once again Europe's considerable ambivalence about its actual "novelty." Like Gastaldi's northern land bridge and Strait of Anian, the visual assimilation of an island traditionally perceived as part of the Orient into the new Occident clearly underscores the fact that those in Europe who truly believed that America was indeed a New World at the same time kept toying with the idea that it might, somehow, still be attached to the Old.

The cartographic display of ambivalence, however, is not limited to the presentation of two conflicting visual statements on the same map. In fact, mapmakers often express their mixed feelings by not making any explicit statement at all.

Consider, for example, how America is portrayed on the Cantino, King-Hamy, and Caveri world maps (Plates 3, 22, and 8). These early maps certainly offered their viewers a lot of information about the new continent's eastern coastline. At the same time, however, they told them absolutely nothing about its interior, not to mention its western bounds, thereby following the traditional format of the fourteenth- and fifteenth-century Italian and Portuguese "portolan" sea charts, which showed Europe's and Africa's coastlines in great detail while revealing nothing about whatever lay beyond them (which was evidently regarded as of little or no interest to sailors). This lack of information clearly reflected the actual state of Europe's geographical knowledge about the New World then. At the same time, however, it also reflected its considerable cosmographic ambivalence about its "novelty."

Quite similar, in fact, was the way in which many European maps kept portraying America's west coast even after Balboa's and Magellan's dramatic discoveries in the Pacific. A perfect example is the way it is portrayed on Juan Vespucci's 1526 world map (Plate 17). With the exception of Central America and southern Chile, the New World's entire western shoreline is still missing, thereby reflecting Europe's absolute ignorance about it at the time. The shaded parts of the map, which represent the land, simply fade out toward the Pacific without offering any definitive statement about the westward extent of the new continent. The contrast between that and the sharp, conclusive delineation of America's eastern coastline is, indeed, quite remarkable.

This glaring contrast is even more striking, in fact, on the Salviati world map from around the same year (Plate 18). Nowhere is the tension between the already discovered and the yet unknown, the "*terra cognita*" and the *terra incognita*, captured more powerfully than on this map. The America portrayed on the Salviati map is a continent clearly bounded (and thus absolutely detached from all other continents) in the east yet still literally open ended (and thus cosmographically ambiguous) in the west.

One of the most distinctive features of all these maps—
and certainly the most explicit visual manifestation of
their producers' cosmographic ambivalence about the New
World—is their tentativeness, so well captured in the way
they depict broken shorelines. As evident from the delineation of the South American coast on the King-Hamy
(Plate 22), Kunstmann II, and Caveri (Plate 8) world
maps and the North and Central American coasts on the
Kunstmann IV map (Plate 31), for example, those broken
shorelines clearly capture the incomplete state of actual
geographical knowledge about America in Europe at the
time. (In the Kunstmann II and King-Hamy maps, not
even the fact that the northern and eastern coasts of Brazil
were actually joined was yet clear to their producers.) At
the same time, however, they also reflect the obvious lack
of readiness to make a definitive, unequivocal cosmographic statement about the New World.

Nowhere is this more explicitly evident than in the way
many mid- and late-sixteenth-century maps deal with the
yet-unexplored Northwest. Thus, for example, in Battista
Agnese's (1542), Sebastian Cabot's (1544), and Edward
Wright's (1599 [Plate 24]) world maps, the cartographers'
ambivalence about the actual cosmographic status of North
America vis-à-vis Asia is quite clear from the way they
rather abruptly discontinued its Pacific shoreline beyond
California. Such a graphic display of ambivalence became
quite prevalent in the late seventeenth century, as many
cartographers were less ready to commit themselves fully
to a cosmographic vision of a New World absolutely detached from the Old, and thus kept using the broken-
shoreline image to portray their still tentative visions of
North America beyond Oregon. Indeed, until the late
eighteenth century, Europe was quite accustomed to the
visual image of an essentially "unfinished" America.

While the cartographers who drew those broken shorelines never claimed that the New World was in fact attached to the Old, they were clearly not ready to dismiss that possibility altogether either. They certainly did not offer their viewers the visual closure they would have needed in order to be absolutely positive that America was indeed fully detached from Asia. By essentially truncating North America, they thus left the question of whether the two were in fact connected to each other literally open!

While still avoiding any definitive statement about its actual cosmographic status relative to Asia, sixteenth-century mapmakers also expressed their ambivalence about the "novelty" of the New World in a somewhat more obtrusive manner than just leaving its shorelines broken. They did so by effectively utilizing the borders of their maps.[94]

In sharp contrast to their spherical counterparts, maps did not force their producers to commit themselves unequivocally to any particular cosmographic vision of the relations between the New World and the Orient. The very fact that they were literally bounded inevitably allowed those who made them to refrain from making any definitive statement about America's situation relative to Asia by simply cutting it off at the edges!

Consider, for example, the la Cosa (1500) world map (Plate 7), Waldseemüller's (1513) *Terre Nove*, or Piri Re'is's (1513) chart of the Ocean Sea. Both North and South America are cut off at the left edge of these maps in a way that clearly prevents their viewers from examining their full westward extent. That is also true of North America on the Cantino (Plate 3) and Caveri (Plate 8) world maps and of South America on Bernard Sylvanus's 1511 world map. By cutting America off at their edges, the producers of these maps clearly avoided having to specify definitively how they believed it was situated relative to Asia, thereby managing to remain noncommittal about its actual cosmographic status. From looking at any of these maps, one can tell absolutely nothing about what happened beyond their left edge and, thus, about how America indeed re lates to Asia. Since the la Cosa, Cantino, and Caveri maps also did not fully close off Asia in the Far East, the possibility that the two continents might actually "meet" somewhere beyond the map's borders could not be ruled out.

While none of these maps actually claims that America is in fact attached to Asia, neither do they state unequivocally that it is not. By cutting off both continents at the maps' edges, their producers certainly managed to leave open the question of whether or not they were connected to each other. (By placing the image of Saint Christopher right where Central America would have been [Plate 7], la Cosa likewise left the yet unresolved nature of the relations between North and South America just as ambiguous.)

The glaring inconclusiveness that characterizes maps that leave America's shorelines broken or simply cut it off at their edges may very well be seen as evidence of downright evasiveness, an intellectual cop-out. At the same time, however, it may also be lauded as a sign of extreme intellectual prudence as well as open-mindedness.

Consider, for example, Jorge Reinel's c.1518 world map. If we look at the map in its entirety, the sheer size of the Pacific makes it hard to believe that Reinel indeed considered it possible that America might somehow be attached to Asia. Yet when we examine closely its American section, commonly known as Kunstmann IV (Plate 31), we are struck by how much of America's west shoreline is actually left tentatively open. Though he obviously did not know about that coast any less than Waldseemüller had eleven years earlier (and before Balboa), Reinel was clearly not ready to make the empirically unsubstantiated intellectual leap that Waldseemüller had made and fully close it off.

We may very well choose to denounce as wishy-washy or intellectually timid those mapmakers who left America's borders open. Yet we may also regard their readiness to forgo the tranquillity usually offered by closure as the mark of a truly "open" mind that is always ready to grapple with ambiguity.[95]

Conclusion

After tracing the long history of our current image of America, it becomes quite clear that it was not discovered by Christopher Columbus on October 12, 1492. The European "discovery" of the New World was not a single event that occurred on a single day but rather a long process that lasted almost three hundred years. The year 1492 was a critical moment in that process but certainly not the only one. In fact, it marks the beginning of a long mental voyage that was fully completed only in the late eighteenth century.

One cannot claim that Europe indeed discovered America in 1492, when its actual image of it at the time was that of a few islands off the shores of China. In order for Europeans fully to discover America, they first had to realize that what Columbus had in fact discovered beyond the Atlantic was a previously unknown fourth continent that was absolutely distinct and separate from the other three, a "New World," so to speak.

That such a continent did not appear all of a sudden full-blown on October 12, 1492, is quite evident from the way it is actually portrayed on European maps from the early sixteenth to the late eighteenth century.[1] As one observer of the sixteenth-century cartographic scene put it, Europe's image of the New World "seems to have developed through a process of slow and painful accretion, with many maps representing abortive efforts to synthesize logical configurations out of fragmentary and confusing information."[2] By vividly documenting how a slowly emerging new continent gradually forced itself into Europe's consciousness, these maps are the best proof that discovering America was indeed a long process rather than a single event. In portraying the glaringly distorted visions of the Western Hemisphere sustained by Europeans long after Columbus, they quite graphically demonstrate that America as we know it was definitely not discovered in 1492.

◆

Part of the reason it took Europe so long fully to discover America was the fact that Columbus's first encounter with it in 1492 actually revealed to his contemporaries only a fractional part of this continent and was evidently insufficient for determining its actual cosmographic status. The full picture of America that we now have could not have possibly been available to anyone back then, as it presupposes, for example, the subsequent discoveries of Vespucci and Magellan in South America, Balboa and Pineda in Central America, Corte-Real and Verrazano in the North Atlantic, and Bering and Cook in the North Pacific. Yet part of the delay was also a result of the fact that the process of discovery presupposes a certain readiness to accept that what one discovers may require changing the way one sees the world.[3] This kind of readiness to challenge the classical tricontinental image of the world was something Columbus and many of his contemporaries obviously did not have.

Furthermore, the mental process of discovering America

was not a simple linear progression from denial through ambivalence to innovation. Those three prototypical responses to the discovery were not an irreversible sequence of phases of a transition from a more conservative to a more progressive cosmography. In fact, actual reversals from avant-garde to more traditional visions of the New World were not at all uncommon in Europe between 1492 and 1778.

A case in point is the early European image of Newfoundland. As early as 1501 it was described by the Portuguese as a land "which never before was known to anyone,"[4] and by 1502 it was actually known in England as "the New Found Land."[5] Nevertheless, five years later, it was still portrayed by Contarini, Ruysch, and Rosselli as part of Asia. Similarly, despite the fact that by 1507 both Pacheco and Waldseemüller had already identified America as a distinct continent, several decades later eminent cartographers like Vopell and Finé were still portraying Mexico and China as essentially one and the same. By the same token, though most late-sixteenth-century European cartographers had unequivocally portrayed America as separated from Asia by a strait, their seventeenth- and eighteenth-century successors were clearly more ambivalent about its actual insularity and very often left its North Pacific shoreline tentatively broken.

Such cosmographic "regressions," in fact, sometimes occurred even on maps by the same individual, as evident from the remarkable case of Martin Waldseemüller himself. Evidently intimidated by his own audacity, the great cosmographer soon "regressed" and in 1513—six years after publishing the first map ever to show America as an absolutely distinct continent and naming it after the man most closely associated with its image as a New World—he contributed to the new edition of Ptolemy's *Geography* a new map (*Terre Nove*) in which, cutting it off at the edge, he basically avoided making any definitive statement about its actual cosmographic status relative to Asia. He also gave up the name *America* and attributed its discovery to Columbus. In his *Carta Marina* three years later, Waldseemüller actually "regressed" even further, to the point of explicitly proclaiming North America part of Asia! (Echoing Columbus, he also identified Hispaniola as the biblical Oriental land of Ophir.[6])

Then there is the very similar case of Johann Schöner. As evident from his 1515 and 1520 (Plate 9) globes, he was one of the very first in Europe to accept Waldseemüller's early cosmographic vision of America and portray it as an island, fully detached from Asia. Yet in his 1533 tract *Opusculum Geographicum*, Schöner suddenly presents Mexico just as it is portrayed by Vopell and Finé, explicitly identifying its capital, Tenochtitlán (Temistitan), as Marco Polo's Chinese counterpart Quinsay: "There is a land called Mexico and Temistitan in Upper India, which in former times was called Quinsay."[7] He then goes on to explain:

> Americus Vespucius, sailing along the coasts of Upper India, from Spain to the west, thought that the said part which is connected with Upper India, was an island which he had caused to be called after his own name. But now other hydrographers of more recent date have found that that land (South America) and others beyond constitute a continent, which is Asia.[8]

He later reiterates his claim even more explicitly: "It has been ascertained that the said country (America) was the continent of Upper India, which is a part of Asia."[9] Contrasting such statements with the "progressive" image of this continent offered by Schöner himself thirteen years earlier (Plate 9) helps remind us why it took Europe so long fully to discover America.

It certainly took more than one person to discover America. Christopher Columbus played a major role in its discovery, but he was clearly not the only one who did. In fact, he was only one of a long list of discoverers of America that, as we have seen, includes a few historical celebrities as well as many unsung heroes. Such a list includes the "visual pioneers" who were the first to see Greenland, North America, the island of Guanahaní, and Alaska. It also includes explorers who landed in America before Columbus as well as others who came there only after him yet whose discoveries were indispensable to the development of our current image of America as a single landmass that is absolutely distinct and separate from Asia. And it also includes the cosmographers, cartographers, and other "intellectual pioneers" who helped Europe understand the cosmographic implications of their discoveries.

In short, the history of the discovery of the New World includes more than a single hero, and though we usually portray Columbus as the one who discovered it, we could just as well have picked Vespucci, Waldseemüller, Balboa, Magellan, or Cook instead. Each of them was just as critical to the development of our present image of America.

In commemorating 1492, we inevitably distort the reality of the process through which America was actually discovered by Europe. By highlighting only Columbus's celebrated landfall in the Bahamas, we relegate practically all the rest of that long and complex process to semioblivion.

We should bear this in mind as we celebrate this year the five-hundredth anniversary of Columbus's first encounter with America. It is one thing to remember Columbus and the day on which he first landed on this side of the Atlantic. It is quite another thing to claim that America was indeed discovered on that day.

As everyone knows, on October 12, 1492, Columbus made a historic landfall on the tiny island of Guanahaní in the Bahamas. America, however, was yet to be discovered.

NOTES

Introduction

1. See, for example, Kirkpatrick Sale, *The Conquest of Paradise* (New York: Alfred A. Knopf, 1990).
2. See, for example, Yael Zerubavel, "The Politics of Interpretation: Tel Hai in Israel's Collective Memory," *AJS Review* 16 (1991):133–159.

Chapter One   Did Columbus Discover America?

1. For some exceptionally convincing accounts of such early contacts between the Old and the New World, see Alexander von Wuthenau, *Unexpected Faces in Ancient America 1500 B.C.–1500 A.D.* (New York: Crown, 1975), and Ivan van Sertima, *They Came Before Columbus* (New York: Random House, 1976). See also Barry Fell, *America B.C.* (Rev. ed. New York: Pocket Books, 1989); Cyrus H. Gordon, *Before Columbus* (New York: Crown, 1971); Constance Irwin, *Fair Gods and Stone Faces* (New York: St. Martin's Press, 1963); Charles M. Boland, *They All Discovered America* (1961; reprint, New York: Pocket Books, 1963); Arlington Mallery and Mary R. Harrison, *The Rediscovery of Lost America* (1951; reprint, New York: E. P. Dutton, 1979); Barry Fell, *Saga America* (New

York: Times Books, 1980); and James Bailey, *The God-Kings and the Titans* (New York: St. Martin's Press, 1973).

2. Armando Cortesão, "The North Atlantic Nautical Chart of 1424," *Imago Mundi* 10 (1953):13; Boland, *They All Discovered America*, pp. 381–393; David B. Quinn, ed., *New American World* (New York: Arno Press & Hector Bye, 1979), vol.1, pp. 79–82, 85–88. But see Samuel E. Morison, *Portuguese Voyages to America in the Fifteenth Century* (Cambridge: Harvard University Press, 1940), pp. 21–51, 119–143.

3. David B. Quinn, "The Argument for the English Discovery of America between 1480 and 1494," *Geographical Journal* 127 (1961):278–279; James A. Williamson, *The Cabot Voyages and Bristol Discovery under Henry VII* (Cambridge: Cambridge University Press, 1962), pp. 19–24, 188–189, 228.

4. Louis-André Vigneras, "New Light on the 1497 Cabot Voyage to America," *Hispanic American Historical Review* 36 (1956):507–509; Vigneras, "The Cape Breton Landfall: 1494 or 1497," *Canadian Historical Review* 38 (1957):226–228. See also David B. Quinn, "John Day and Columbus," *Geographical Journal* 133 (1967):205–209; Samuel E. Morison, *The European Discovery of America—The Northern Voyages A.D. 500–1600* (New York: Oxford University Press, 1971), pp. 205–206.

5. Quinn, "The Argument for the English Discovery of America," p. 282.

6. Ibid.; Williamson, *The Cabot Voyages*, p. 46.

7. Birgitta L. Wallace, "The L'Anse aux Meadows Site," in Gwyn Jones, *The Norse Atlantic Saga* (Oxford: Oxford University Press, 1986), pp. 297–298.

8. Magnus Magnusson and Hermann Pálsson, trans., *The Vinland Sagas* (Harmondsworth: Penguin, 1965), pp. 24–25; Gwyn Jones, "The Western Voyages and the Vinland Map," in Wilcomb E. Washburn, ed., *Proceedings of the Vinland Map Conference* (Chicago: University of Chicago Press, 1971), p. 122.

9. Magnusson and Pálsson, *The Vinland Sagas*, pp. 26–27;

Helge Ingstad, *Westward to Vinland* (New York: St. Martin's Press, 1969), p. 29; Gwyn Jones, *The Norse Atlantic Saga* (Oxford: Oxford University Press, 1986), pp. 148, 304.

10. Magnusson and Pálsson, *The Vinland Sagas,* pp. 24, 49–72, 75–105; Jones, *The Norse Atlantic Saga,* pp. 121, 186–232, 306–307.

11. Magnusson and Pálsson, *The Vinland Sagas,* pp. 15–16, 24.

12. Ibid., p. 28; R. A. Skelton, "The Vinland Map," in R. A. Skelton et al., *The Vinland Map and the Tartar Relation* (New Haven: Yale University Press, 1965), pp. 140–141, 224–226; Jones, "The Western Voyages and the Vinland Map," pp. 122–123.

13. Magnusson and Pálsson, *The Vinland Sagas,* p. 15; Jones, *The Norse Atlantic Saga,* p. 20.

14. Louis-André Vigneras, review of *The Vinland Map and the Tartar Relation,* by R. A. Skelton et al., *English Historical Review* 83 (1968):118.

15. Skelton, "The Vinland Map," p. 226n.

16. Jones, *The Norse Atlantic Saga,* pp. 20–22.

17. See also Ingstad, *Westward to Vinland,* pp. 85–87.

18. This name is also used in the 1623 *Groenlands annál* as another name for Helluland. See Jones, *The Norse Atlantic Saga,* pp. 20–22.

19. Magnusson and Pálsson, *The Vinland Sagas,* p. 120.

20. Ingstad, *Westward to Vinland;* Wallace, "The L'Anse aux Meadows Site."

21. Magnusson and Pálsson, *The Vinland Sagas,* p. 28; Jones, "The Western Voyages and the Vinland Map," p. 125.

22. Ingstad, *Westward to Vinland,* pp. 92–93.

23. William Robertson, *The History of America* (London: Strahan, Cadell, and Balfour, 1777), vol. 1, pp. 438–439; Johann R. Forster, *History of the Voyages and Discoveries Made in the North* (Dublin: 1786), pp. 83, 86, 204, 439n.

24. Boland, *They All Discovered America,* pp. 153–155, 173, 236–237; Morison, *The European Discovery of America—The Northern Voyages,* p. 37.

25. See, for example, Boland, *They All Discovered America*, pp. 193–295, 339–365; Fell, *Saga America*, pp. 341–370.

26. Erik Wahlgren, *The Kensington Stone—A Mystery Solved* (Madison: University of Wisconsin Press, 1958).

27. Ibid., p. 202.

28. Ibid., pp. 122, 180; A. W. Brøgger and Haakon Shetelig, *The Viking Ships* (Oslo: Dreyers Verlag, 1951), pp. 141–142; Morison, *The European Discovery of America—The Northern Voyages*, p. 38.

29. "Goodbye Columbus, Hello Leif," *New York Times*, July 26, 1991, p. A26.

30. Sale, *The Conquest of Paradise*, p. 350.

31. Vilhjalmur Stefansson, *Ultima Thule* (New York: Macmillan, 1940), p. 172.

32. Morison, *Portuguese Voyages to America*; Cortesão, "The North Atlantic Nautical Chart of 1424."

33. Stefansson, *Ultima Thule*, pp. 111–112, 211. See also Alwyn A. Ruddock, "Columbus and Iceland: New Light on An Old Problem," *Geographical Journal* 136 (1970):179.

34. R. A. Skelton et al., *The Vinland Map and the Tartar Relation* (New Haven: Yale University Press, 1965).

35. "Yale Press Publishes 'Map Discovery of the Century,'" *Publishers Weekly*, October 11, 1965, p. 34.

36. The publication of the Vinland Map was featured, along with the map itself, on the front page of the *New York Times*.

37. See, for example, the lists of the reviews in Wilcomb E. Washburn, ed., *Proceedings of the Vinland Map Conference* (Chicago: University of Chicago Press, 1971), pp. 155–181.

38. *New York Times*, October 13, 1965, p. 27; *The Times* (London), October 13, 1965, p. 10.

39. *New York Times*, October 12, 1965, p. 1; *The Times*, October 13, 1965, p. 10.

40. *New York Times*, October 13, 1965, p. 27.

41. Ibid., October 12, 1965, p. 58.

42. See, for example, Yael Zerubavel, *The Politics of Commemoration* (forthcoming).

43. Van Sertima, *They Came Before Columbus.*
44. *Newsweek*, September 23, 1991, p. 50.
45. See, for example, Boland, *They All Discovered America.*
46. On the conventional nature of delineating socially meaningful mental clusters, see Eviatar Zerubavel, *The Fine Line* (New York: Free Press, 1991), pp. 61–80.
47. Williamson, *The Cabot Voyages*, p. 208.
48. On his return to England in August 1497, Cabot apparently displayed a map as well as a globe showing his discoveries beyond the Atlantic. Copies of that map were evidently sent soon after that to Columbus himself (by English merchant John Day) and to Ferdinand and Isabella (by the Spanish ambassador to England, Pedro de Ayala) and most probably found their way to la Cosa. See Vigneras, "The Cape Breton Landfall," pp. 222–223; Williamson, ibid., pp. 209, 212; David B. Quinn, *England and the Discovery of America, 1481–1620* (New York: Alfred A. Knopf, 1974), p. 106; Quinn, *The New American World*, vol. 1, pp. 101–102.
49. Arthur Davies, "The 'English' Coasts on the Map of Juan de la Cosa," *Imago Mundi* 13 (1956):26–29; Williamson, *The Cabot Voyages*, pp. 72–83; W. F. Ganong, *Crucial Maps in the Early Cartography and Place-Nomenclature of the Atlantic Coast of Canada* (Toronto: University of Toronto Press, 1964), pp. 8–39; Arthur Davies, "The Date of Juan de la Cosa's World Map and Its Implications for American Discovery," *Geographical Journal* 142 (1976):111–116.
50. Williamson, *The Cabot Voyages*, pp. 56, 63. See also Vigneras, "New Light on the 1497 Cabot Voyage to America," p. 503; Vigneras, "The Cape Breton Landfall," p. 227.
51. Jones, *The Norse Atlantic Saga*, p. 171. See also Magnusson and Pálsson, *The Vinland Sagas*, p. 17.
52. Ibid., pp. 49–51, 76ff.
53. Zerubavel, *The Fine Line*, pp. 61–80.
54. Ibid., pp. 24ff.
55. See, however, Washburn, *Proceedings of the Vinland Map Conference*, p. x; Raymond H. Ramsay, *No Longer on the Map*

(New York: Ballantine, 1972), p. 152; Barbara J. Sivertsen, letter to the editor, *New York Times*, December 12, 1990, p. A22.

56. On the history of that, see Sale, *The Conquest of Paradise*, pp. 333–362.

57. Y. Zerubavel, *The Politics of Commemoration*.

58. Samuel E. Morison, *The European Discovery of America—The Southern Voyages* A.D. *1492–1616* (New York: Oxford University Press, 1974), p. 203.

59. Ibid.; Charles L. G. Anderson, *Life and Letters of Vasco Núñez de Balboa* (Westport, Conn.: Greenwood, 1941), pp. 174–175.

60. Magnusson and Pálsson, *The Vinland Sagas*, pp. 16–17; Jones, *The Norse Atlantic Saga*, pp. 73–74.

61. Magnusson and Pálsson, *The Vinland Sagas*, pp. 51–54.

62. Ibid., p. 55; Jones, *The Norse Atlantic Saga*, pp. 115–116. But see also Ingstad, *Westward to Vinland*, p. 60.

63. Magnusson and Pálsson, *The Vinland Sagas*, pp. 50, 77; Jones, *The Norse Atlantic Saga*, pp. 186, 209.

64. Magnusson and Pálsson, *The Vinland Sagas*, pp. 54–55.

65. Ferdinand Columbus, *The Life of the Admiral Christopher Columbus* (before 1539; reprint, New Brunswick, N.J.: Rutgers University Press, 1959), pp. 58–59; Henry Harrisse, *The Discovery of North America* (1892; reprint, Amsterdam: N. Israel, 1969), p. 665; Robert H. Fuson, trans., *The Log of Christopher Columbus* (Camden, Maine: International Marine Publishing Co., 1987), p. 75; Oliver Dunn and James E. Kelley, eds., *The Diario of Christopher Columbus's First Voyage to America 1492–1493* (Norman: University of Oklahoma Press, 1989), p. 59; Sale, *The Conquest of Paradise*, pp. 62–63, 399n.

66. For various attempts to change that, however, see Bartolomé de Las Casas, *History of the Indies* (c. 1550–1565; reprint, New York: Harper Torchbooks, 1971), p. 62; Sale, *The Conquest of Paradise*, pp. 216, 335, 340.

67. See, for example, Morison, *The European Discovery of America—The Southern Voyages*, pp. 276ff.

68. See also Stefan Zweig, *Amerigo* (New York: Viking, 1942); Frederick J. Pohl, *Amerigo Vespucci—Pilot Major* (New York: Columbia University Press, 1944), pp. 176–177.

69. See also Edmundo O'Gorman, *The Invention of America* (Bloomington: Indiana University Press, 1961).

Chapter Two   The Mental Discovery of America

1. Cecil Jane, ed., *The Four Voyages of Columbus* (1930; reprint, New York: Dover, 1988), vol. 1, pp. cxxiii-cxliii; Rudolf Hirsch, "Printed Reports on the Early Discoveries and Their Reception," in Fredi Chiappelli, ed., *First Images of America* (Berkeley: University of California Press, 1976), vol. 2, p. 553; Kenneth Nebenzahl, *Atlas of Columbus and the Great Discoveries* (Chicago: Rand McNally, 1990), p. 28.

2. O'Gorman, *The Invention of America*, p. 124.

3. Williamson, *The Cabot Voyages*, pp. 207–209.

4. Rodney W. Shirley, *The Mapping of the World* (1984; reprint, London: The Holland Press, 1987), p. 97.

5. See also Zerubavel, *The Fine Line*, p. 76.

6. Ibid., p. 79.

7. Quinn, *New American World*, vol. 1, p. 150. See also Harrisse, *The Discovery of North America*, pp. 108–109; George H. T. Kimble, "Introduction" to Duarte Pacheco Pereira, *Esmeraldo de Situ Orbis* (London: Hakluyt Society [2nd series, no. 79], 1937), p. xxiii.

8. Kimble, ibid., pp. xvi-xvii.

9. Ibid., pp. xxiii-xxiv.

10. Duarte Pacheco Pereira, *Esmeraldo de Situ Orbis* (1505– 1508; reprint, London: Hakluyt Society [2nd series, no. 79], 1937), p. 12.

11. Williamson, *The Cabot Voyages*, pp. 88–89, 210.

12. Ibid., pp. 108–110, 234; Harrisse, *The Discovery of North America*, pp. 43–44, 116–117; David O. True, "Some Early Maps Relating to Florida," *Imago Mundi* 11 (1954):73–84;

Lawrence C. Wroth, *The Voyages of Giovanni da Verrazzano 1524–1528* (New Haven: Yale University Press, 1970), pp. 40–43; John H. Parry and Robert G. Keith, eds., *New Iberian World* (New York: Times Books and Hector & Rose, 1984), vol. 2, p. 378.

13. Louis-André Vigneras, *The Discovery of South America and the Andalusian Voyages* (Chicago: University of Chicago Press, 1976), pp. 47–63, 69–94, 99–109. See also his maps nos. 5, 6, 8, 9, and 10 as well as Felipe Fernández-Armesto, ed., *The Times Atlas of World Exploration* (New York: HarperCollins, 1991), pp. 51, 68–69.

14. Vigneras, *The Discovery of South America,* p. 105; Parry and Keith, *New Iberian World,* vol. 3, p. 1.

15. Pohl, *Amerigo Vespucci,* p. 131. Some Vespucci scholars, most notably Roberto Levillier (see his *America la Bien Llamada* [Buenos Aires: Guillermo Kraft, 1948] and "New Light on Vespucci's Third Voyage: Evidence of His Route and Landfalls," *Imago Mundi* 11 [1954]:37–46), accept Vespucci's claim that he did indeed reach that far south. So did Johannes Ruysch, as evident from one of the inscriptions on his 1507 world map.

16. Parry and Keith, *New Iberian World,* vol. 3, pp. 130–132; Fernández-Armesto, *Times Atlas of World Exploration,* p. 52.

17. Morison, *The European Discovery of America—The Southern Voyages,* pp. 517–518.

18. See also Peter Martyr's 1511 map of the Indies.

19. See also Piri Re'is's 1513 chart of the "Ocean Sea."

20. Quoted in Wroth, *The Voyages of Giovanni da Verrazzano,* pp. 142–143. Italics mine.

21. Quoted in ibid., p. 142.

22. Ibid., p. 89.

23. Zerubavel, *The Fine Line,* pp. 1, 21, 118.

24. John H. Parry, *The Discovery of South America* (New York: Taplinger Publishing Co., 1979), p. 120.

25. See also Eviatar Zerubavel, "The Standardization of Time: A Sociohistorical Perspective," *American Journal of Sociology* 88 (1982):7–8.

26. Derek Howse, *Greenwich Time and the Discovery of the Longitude* (Oxford: Oxford University Press, 1980).

27. Morison, *The European Discovery of America—The Southern Voyages*, pp. 406–409; Fernández-Armesto, *Times Atlas of World Exploration*, pp. 164–165.

28. See also Zerubavel, "The Standardization of Time," pp. 15–16.

29. Henry R. Wagner, *Spanish Voyages to the Northwest Coast of America in the Sixteenth Century* (San Francisco: California Historical Society, 1929), pp. 2–4.

30. Ibid., pp. 6, 14.

31. Ibid., pp. 74, 92, 155, 337; Warren L. Cook, *Flood Tide of Empire* (New Haven: Yale University Press, 1973), p. 8; Fernández-Armesto, *Times Atlas of World Exploration*, p. 103.

32. Wagner, *Spanish Voyages to the Northwest Coast of America*, pp. 253–255, 263.

33. Morison, *The European Discovery of America—The Northern Voyages*, p. 506.

34. Justin Winsor, *Narrative and Critical History of America* (Boston: Houghton Mifflin & Co., 1886), vol. 2, p. 439.

35. C. W. Butterfield, *History of the Discovery of the Northwest by John Nicolet in 1634* (1881; reprint, Port Washington, N.Y.: Kennikat Press, 1969), pp. 58–59, 102.

36. Johann G. Kohl, "Asia and America," *Proceedings of the American Antiquarian Society* 21 (1911):312; L. Breitfuss, "Early Maps of North-Eastern Asia and of the Lands around the North Pacific," *Imago Mundi* 3 (1939):87; Eric Newby, *World Atlas of Exploration* (1975; reprint, New York: Crescent, 1985), pp. 146–147; Piers Pennington, *The Great Explorers* (1979; reprint, London: Bloomsbury, 1989), p. 160; Raymond H. Fisher, *The Voyage of Semen Dezhnev in 1648* (London: Hakluyt Society [2nd series, vol. 159], 1981); Fernández-Armesto, *Times Atlas of World Exploration*, p. 159.

37. Fisher, *The Voyage of Semen Dezhnev*, p. 273. See also pp. 275–276.

38. Ibid., p. 275. Leo Bagrow, "The First Russian Maps of Si-

beria and their Influence on the West-European Cartography of N. E. Asia," *Imago Mundi* 9 (1952):83–93.

39. See also Raymond H. Fisher, *Bering's Voyages* (Seattle: University of Washington Press, 1977), pp. 33–37.

40. Ibid., pp. 12, 23.

41. Ibid., p. 9; Frank A. Golder, *Bering's Voyages* (New York: American Geographical Society, 1922), vol. 1, p. 4.

42. Gerhard F. Müller, *Bering's Voyages—the Reports from Russia* (1758; reprint, Fairbanks: University of Alaska Press, 1986), p. 17. See also Georg W. Steller, *Journal of a Voyage with Bering 1741–1742* (1743; reprint, Stanford: Stanford University Press, 1988), p. 47; Golder, *Bering's Voyages*, vol. 1, p. 11; Frank A. Golder, *Russian Expansion on the Pacific, 1641–1850* (Cleveland: Arthur H. Clark, 1914), pp. 91, 114, 134; Fisher, *Bering's Voyages*, p. 12.

43. Wagner, *Spanish Voyages to the Northwest Coast of America*, pp. 113–116; Morison, *The European Discovery of America—The Southern Voyages*, pp. 494–495; Fernández-Armesto, *Times Atlas of World Exploration*, p. 164.

44. It is quite possible that the same feat had actually been accomplished a few months earlier by Alonso de Arellano. See Wagner, *Spanish Voyages to the Northwest Coast of America*, pp. 111–112, 118.

45. See Golder, *Bering's Voyages*, vol. 1, p. 20.

46. Ibid., pp. 29–30; Golder, *Russian Expansion on the Pacific*, pp. 169–170; Fisher, *Bering's Voyages*, pp. 143–144.

47. Golder, *Russian Expansion on the Pacific*, pp. 161–163; Golder, *Bering's Voyages*, vol. 1, pp. 22–24; Breitfuss, "Early Maps of North-Eastern Asia," p. 89.

48. Fisher, *Bering's Voyages*, p. 38.

49. Ibid., p. 169; Glyndwr Williams, "Alaska Revealed: Cook's Explorations in 1778," in Antoinette Shalkop, ed., *Exploration in Alaska* (Anchorage: Cook Inlet Historical Society, 1980), p. 69.

50. Steller, *Journal of a Voyage with Bering*, p. 60; Müller, *Bering's Voyages*, p. 101; Golder, *Bering's Voyages*, vol. 1, pp. 93,

290–291, 313, 332, 342; Fernández-Armesto, *Times Atlas of World Exploration*, p. 104.

51. Golder, *Bering's Voyages*, vol. 1, pp. 291, 313.
52. Steller, *Journal of a Voyage with Bering*, p. 47.
53. Ibid., pp. 72, 97–107; Müller, *Bering's Voyages*, pp. 102, 104, 107, 109–110; Golder, *Bering's Voyages*, vol. 1, pp. 274–275.
54. Lawrence C. Wroth, "The Early Cartography of the Pacific," *Papers of the Bibliographical Society of America* 38 (1944): 222–224.
55. See, for example, Müller, *Bering's Voyages*, pp. 43, 53, 137–138; Henry R. Wagner, *The Cartography of the Northwest Coast of America to the Year 1800* (Berkeley: University of California Press, 1937), vol. 1, p. 156; Williams, "Alaska Revealed," p. 69.
56. Stuart R. Tompkins and Max L. Moorhead, "Russia's Approach to America," *British Columbia Historical Quarterly* 13 (1949):233–249; Cook, *Flood Tide of Empire*, pp. 44–46, 54; Christon I. Archer, "Russians, Indians, and Passages: Spanish Voyages to Alaska in the Eighteenth Century," in Antoinette Shalkop, ed., *Exploration in Alaska* (Anchorage: Cook Inlet Historical Society, 1980), pp. 129, 142; Glynn Barratt, *Russia in Pacific Waters 1715–1825* (Vancouver: University of British Columbia Press, 1981), pp. 51–53, 67–68. For a list of documented Russian expeditions to Alaska between Bering and Chirikov's 1741 voyage and Pérez's 1774 voyage, see Stuart R. Tompkins, "After Bering: Mapping the North Pacific," *British Columbia Historical Quarterly* 19 (1955):11–24.
57. Herbert K. Beals, *Juan Pérez on the Northwest Coast* (Portland: Oregon Historical Society, 1989), pp. 12–13, 39, 76, 126, 140–143, 201–203, 236.
58. Francisco A. Mourelle, *Voyage of the Sonora in the Second Bucareli Expedition* (1781; reprint, San Francisco: Thomas C. Russell, 1920), pp. 47, 71; Wagner, *The Cartography of the Northwest Coast of America*, vol. 1, pp. 176–178; Cook, *Flood Tide of Empire*, p. 81; Herbert K. Beals, *For Honor and Country* (Portland: Oregon Historical Society, 1985), p. 37.

59. Williams, "Alaska Revealed," p. 69.
60. Wagner, *The Cartography of the Northwest Coast of America*, vol. 1, p. 184; Cook, *Flood Tide of Empire*, pp. 85–86; Beals, *For Honor and Country*, pp. 40, 138.
61. Wagner, *The Cartography of the Northwest Coast of America*, pp. 185–188; J. C. Beaglehole, ed., *The Voyage of the Resolution and Discovery 1776–1780* (London: Cambridge University Press, 1967); Fernández-Armesto, *Times Atlas of World Exploration*, p. 104.
62. Wagner, *The Cartography of the Northwest Coast of America*, p.188.
63. Beaglehole, *The Voyage of the Resolution and Discovery*, part 1, pp. 414, 430, 433–435, 451, 456.

Chapter Three   *The Psychology of Discovering America*

1. Jane, *The Four Voyages of Columbus*, vol.1, pp. 2, 4, 10, 12, 14, 16.
2. Armando Cortesão, *History of Portuguese Cartography* (Coimbra, Portugal: Junta de Investigações do Ultramar, 1969), vol.1, pp. 215–232; Tony Campbell, "Portolan Charts from the Late Thirteenth Century to 1500," in J. B. Harley and David Woodward, eds., *The History of Cartography* (Chicago: University of Chicago Press, 1987), vol.1, pp. 371–463.
3. Zerubavel, *The Fine Line*, p. 119.
4. See also Pierluigi Portinaro and Franco Knirsch, *The Cartography of North America 1500–1800* (New York: Crescent, 1987), p. 35; Nebenzahl, *Atlas of Columbus*, p. 26.
5. See, for example, Ludwik Fleck, *Genesis and Development of a Scientific Fact* (1935; reprint, Chicago: University of Chicago Press, 1981); Karl Mannheim, *Ideology and Utopia* (New York: Harvest, 1936); Tamotsu Shibutani, "Reference Groups as Perspectives," *American Journal of Sociology* 60 (1955):562–569.
6. Fleck, *Genesis and Development of a Scientific Fact*.

7. Ibid.; Thomas S. Kuhn, *The Structure of Scientific Revolutions* (1962; reprint, Chicago: University of Chicago Press, 1970); Eviatar Zerubavel, "If Simmel Were a Fieldworker: On Formal Sociological Theory and Analytical Field Research," *Symbolic Interaction* 3 (1980), 2:25–33.

8. John B. Thacher, *Christopher Columbus* (New York: G. P. Putnam's Sons, 1903), vol. 1, p. 59.

9. Vigneras, *The Discovery of South America*, pp. 3–4.

10. Thacher, *Christopher Columbus*, vol. 1, p. 60; Arthur P. Newton, "Asia or Mundus Novus?" in Arthur P. Newton, ed., *The Great Age of Discovery* (1932; reprint, New York: Burt Franklin, 1970), p. 106; Pohl, *Amerigo Vespucci*, p. 228; O'Gorman, *The Invention of America*, p. 84; Morison, *The European Discovery of America—The Southern Voyages*, pp. 98–99; Parry, *The Discovery of South America*, p. 76.

11. Thacher, *Christopher Columbus*, vol. 1, p. 62.

12. Williamson, *The Cabot Voyages*, p. 214.

13. Quinn, *New American World*, vol. 1, p. 150.

14. Williamson, *The Cabot Voyages*, p. 106; Quinn, *England and the Discovery of America*, pp. 129–130.

15. Williamson, *The Cabot Voyages*, pp. 106, 143–144, 216; Quinn, *England and the Discovery of America*, p. 10.

16. Williamson, *The Cabot Voyages*, pp. 106, 135n; Harrisse, *The Discovery of North America*, pp. 102ff.

17. Williamson, *The Cabot Voyages*, p. 209.

18. Zerubavel, *The Fine Line*, pp. 6–9.

19. Y. Zerubavel, *The Politics of Commemoration*. On the discontinuous perception of time, see also Eviatar Zerubavel, *The Fine Line*, pp. 9–10; Eviatar Zerubavel, *Patterns of Time in Hospital Life* (Chicago: University of Chicago Press, 1979), pp. 5–34; Eviatar Zerubavel, *The Seven-Day Circle* (New York: Free Press, 1985), pp. 121–129.

20. Zerubavel, *The Fine Line*, pp. 24–32.

21. Thacher, *Christopher Columbus*, vol. 1, p. 65.

22. O'Gorman, *The Invention of America*, p. 106; Morison, *The European Discovery of America—The Southern Voyages*, pp. 276ff;

Parry, *The Discovery of South America*, p. 91. But see also Germán Arciniegas, *Amerigo and the New World* (New York: Alfred A. Knopf, 1955), p. 217.

23. Pohl, *Amerigo Vespucci*, p. 130.

24. Parry and Keith, *New Iberian World*, vol. 5, pp. 18–19.

25. Hirsch, "Printed Reports on the Early Discoveries and Their Reception," pp. 547–548, 554–556.

26. Pohl, *Amerigo Vespucci*, pp. 158ff.

27. David B. Quinn, "New Geographical Horizons," in Fredi Chiappelli, ed., *First Images of America* (Berkeley: University of California Press, 1976), vol. 2, pp. 639–647.

28. On the semantics of the terminology used around the discovery of America, see Wilcomb E. Washburn, "The Meaning of 'Discovery' in the Fifteenth and Sixteenth Centuries," *American Historical Review* 68 (1962):1–21.

29. Pacheco, *Esmeraldo de Situ Orbis*, pp. 14–15.

30. Vigneras, *The Discovery of South America*, p. 143.

31. Kimble, "Introduction," pp. xxix–xxx.

32. On the possible Portuguese roots of Waldseemüller's cosmographic vision of America, see also Edward L. Stevenson, *Terrestrial and Celestial Globes* (New Haven: Yale University Press, 1921), vol. 1, pp. 68–69.

33. Carl O. Sauer, *Sixteenth Century North America* (Berkeley: University of California Press, 1971), pp. 15–16.

34. Martin Waldseemüller, *Cosmographiae Introductio* (1507; reprint, Ann Arbor, Mich.: University Microfilms Inc., 1966), p. 70. See also p. 68.

35. Ibid., p. 70.

36. O'Gorman, *The Invention of America*, p. 131.

37. Waldseemüller, *Cosmographiae Introductio*, p. 70.

38. Stevenson, *Terrestrial and Celestial Globes*, vol. 1, p. 71; Newton, "Asia or Mundus Novus?" p. 118; Hirsch, "Printed Reports on the Early Discoveries and Their Reception," p. 556; Daniel J. Boorstin, *The Discoverers* (New York: Random House, 1983), p. 253.

39. See also Edward Heawood, "Glareanus: His Geography and Maps," *Geographical Journal* 25 (1905):652.

40. J. Enterline, "The Southern Continent and the False Strait of Magellan," *Imago Mundi* 26 (1972):54.

41. Waldseemüller, *Cosmographiae Introductio*, p. 78.

42. Zerubavel, *The Fine Line*, pp. 117, 121.

43. Ibid., pp. 6, 35, 47; Suzanne J. Kessler and Wendy McKenna, *Gender* (New York: John Wiley, 1978), pp. 142–153. See also Else Frenkel-Brunswik, "Intolerance of Ambiguity as An Emotional and Perceptual Personality Variable," *Journal of Personality* 18 (1949):128; Harold Garfinkel, *Studies in Ethnomethodology* (Englewood Cliffs, N.J.: Prentice-Hall, 1967), pp. 79–94; Carolyn M. Bloomer, *Principles of Visual Perception* (New York: Van Nostrand Reinhold, 1976), p. 49; Mary W. Helms, *Ulysses' Sail* (Princeton: Princeton University Press, 1988), pp. 172–182.

44. Kuhn, *The Structure of Scientific Revolutions*. On "monster adjustment" in science, see Imre Lakatos, *Proofs and Refutations* (Cambridge: Cambridge University Press, 1976), p. 30; David Bloor, "Polyhedra and the Abominations of Leviticus: Cognitive Styles in Mathematics," in Mary Douglas, ed., *Essays in the Sociology of Perception* (London: Routledge & Kegan Paul, 1982), p. 200.

45. Peter Martyr, *The Decades of the New World or West India*, in Edward Arber, ed., *The First Three English Books on America* (1511–1555; reprint, Birmingham: 1885), p. 98.

46. Parry and Keith, *New Iberian World*, vol. 2, p. 158; Vigneras, *The Discovery of South America*, p. 49.

47. E.G.R. Taylor, "Idée Fixe: The Mind of Christopher Columbus," *Hispanic American Historical Review* 11 (1931):298.

48. Dunn and Kelley, *The Diario of Christopher Columbus's First Voyage to America*, pp. 63, 65.

49. Marco Polo, *The Travels of Marco Polo the Venetian* (c. 1300; reprint, New York: Boni & Liveright, 1926), p. 268.

50. Dunn and Kelley, *The Diario of Christopher Columbus's First Voyage to America*, pp. 73, 109, 111, 113, 115, 119, 125, 129, 217, 273, 285, 307.

51. Ibid., p. 129.

52. Ibid., pp. 273, 285. See also pp. 111, 113, 115.
53. Ibid., pp. 107, 205; Jane, *The Four Voyages of Columbus*, vol. 1, p. 16; Quinn, "New Geographical Horizons," pp. 637–638.
54. Dunn and Kelley, *The Diario of Christopher Columbus's First Voyage to America*, pp. 129, 131.
55. Martyr, *The Decades of the New World*, p. 75; Jane, *The Four Voyages of Columbus*, vol. 1, p. 118.
56. Jane, *The Four Voyages of Columbus*, vol. 1, pp. 114–116.
57. Ibid., p. 120.
58. Martyr, *The Decades of the New World*, p. 75; Ferdinand Columbus, *The Life of the Admiral Christopher Columbus*, p. 75; Taylor, "Idée Fixe," pp. 293–294.
59. Thacher, *Christopher Columbus*, vol. 2, pp. 321, 328.
60. Parry and Keith, *New Iberian World*, vol. 2, p. 99. Italics mine.
61. Ibid., pp. 99–100. Italics mine.
62. Genesis 2:8; Jane, *The Four Voyages of Columbus*, vol. 2, pp. 35n–37n.
63. Parry and Keith, *New Iberian World*, vol. 2, p. 101.
64. Jane, *The Four Voyages of Columbus*, vol. 2, pp. 30, 32, 36, 38, 42.
65. See also George E. Nunn, *The Geographical Conceptions of Columbus* (New York: American Geographical Society, 1924), pp. 71–77.
66. Vigneras, *The Discovery of South America*, p. 105.
67. Ibid., p. 107; Parry and Keith, *New Iberian World*, vol. 3, p. 1.
68. Jane, *The Four Voyages of Columbus*, vol. 2, p. 104. See also pp. 6–7.
69. Martyr, *The Decades of the New World*, p. 66.
70. Jane, *The Four Voyages of Columbus*, vol. 2, pp. 80–82.
71. Ibid., pp. 82–84.
72. Boorstin, *The Discoverers*, pp. 229–230. See also Jane, *The Four Voyages of Columbus*, vol. 2, p. 40; W.G.L. Randles, "The Evaluation of Columbus's 'India' Project by Portuguese and Spanish Cosmographers in the Light of the Geo-

graphical Science of the Period," *Imago Mundi* 42 (1990): 50–51.

73. Jane, *The Four Voyages of Columbus,* vol. 2, p. 84.

74. Donald Weinstein, *Ambassador from Venice* (Minneapolis: University of Minnesota Press, 1960), pp. 61–62, 100.

75. See, for example, Morison, *The European Discovery of America—The Southern Voyages,* p. 246.

76. Nunn, *The Geographical Conceptions of Columbus,* pp. 65–69; George E. Nunn, "The Three Maplets Attributed to Bartholomew Columbus," *Imago Mundi* 9 (1952):16–17, 20–22; Nebenzahl, *Atlas of Columbus,* pp. 38–39.

77. John Bigelow, "The So-Called Bartholomew Columbus Map of 1506," *Geographical Review* 25 (1935):655; Wilcomb E. Washburn, "Japan on Early European Maps," *Pacific Historical Review* 21 (1952):232.

78. See, for example, the Nancy globe and Van den Putte's 1570 copy of Vopell's 1545 world map.

79. See also Finé's 1531 world map.

80. See, for example, Van den Putte's 1570 copy of Vopell's 1545 world map.

81. See also ibid.

82. See, for example, ibid., Georg Braun's 1574 world map, Mario Cartaro's 1579 astronomical diagrams, and Giacomo Franco's 1587 world map.

83. Wagner, *Spanish Voyages to the Northwest Coast of America,* p. 8; A. Grove Day, *Coronado's Quest* (Berkeley: University of California Press, 1940), pp. 64, 334.

84. Quinn, *New American World,* vol. 1, p. 390. See also p. 396.

85. Frederick W. Hodge, ed., *Spanish Explorers in the Southern United States 1528–1543* (1907; reprint, New York: Barnes & Noble, 1946), p. 360.

86. Quinn, *New American World,* vol. 1, p. 412.

87. Shirley, *The Mapping of the World,* pp. 121, 126–127.

88. Margaret S. Mahler et al., *The Psychological Birth of the Human Infant* (New York: Basic Books, 1975); Zerubavel, *The Fine Line,* pp. 14, 118.

89. See also Lloyd Silverman et al., *The Search for Oneness* (New York: International Universities Press, 1982); Richard A. Koenigsberg, *Symbiosis and Separation* (New York: Library of Art and Social Science, 1989), pp. 25–31; Zerubavel, *The Fine Line*, pp. 86ff.

90. D. W. Winnicott, "Transitional Objects and Transitional Phenomena," in *Playing and Reality* (London: Tavistock, 1971), pp. 1–25.

91. Portinaro and Knirsch, *The Cartography of North America*, p. 284.

92. Waldseemüller, *Cosmographiae Introductio*, p. 76. Italics mine.

93. Washburn, "Japan on Early European Maps," p. 225; Nebenzahl, *Atlas of Columbus*, p. 19.

94. On the strategic manipulation of boundaries, see Y. Zerubavel, *The Politics of Commemoration*.

95. See, for example, Zerubavel, *The Fine Line*, pp. 118–122.

Conclusion

1. For an excellent visual "tour" of European cartography during some of that period, see Shirley, *The Mapping of the World*. See also Portinaro and Knirsch, *The Cartography of North America 1500–1800*; Nebenzahl, *Atlas of Columbus and the Great Discoveries*.

2. Bernard G. Hoffman, *Cabot to Cartier* (Toronto: University of Toronto Press, 1961), p. 101.

3. See, for example, Kuhn, *The Structure of Scientific Revolutions*.

4. Quinn, *New American World*, vol. 1, p. 150.

5. Williamson, *The Cabot Voyages*, pp. 106, 143–144; Quinn, *England and the Discovery of America*, p. 110.

6. Washburn, "Japan on Early European Maps," p. 229.

7. Stevenson, *Terrestrial and Celestial Globes*, vol. 1, p. 109. See also Winsor, *Narrative and Critical History of America*, vol. 2, p. 432; Washburn, "Japan on Early European Maps," p. 232.

8. Stevenson, *Terrestrial and Celestial Globes*, vol. 1, p. 109.

9. Ibid., p. 110.

LIST OF MAPS

* c. 1506. Alessandro Zorzi Sketch Map. Italy. Reproduced in plate 21.

* 1506. Giovanni Matteo Contarini World Map. Florence. Reproduced in plate 14.

* 1507. Johannes Ruysch World Map. Rome. Reproduced in plate 2.

* 1507. Martin Waldseemüller World Map. Strassburg. Reproduced in plates 5 and 6.

1507. Martin Waldseemüller Globe Gores. Strassburg. Reproduced in Nebenzahl, *Atlas of Columbus and the Great Discoveries,* p. 27.

c. 1508. Francesco Rosselli Oval Projection World Map. Florence. Reproduced in Nebenzahl, *Atlas of Columbus and the Great Discoveries,* p. 57.

* c. 1508. Francesco Rosselli Marine Chart of the World. Florence. Reproduced in plate 13.

c. 1510. "Lenox (Hunt)" Globe. Reproduced in Stevenson, *Terrestrial and Celestial Globes,* vol. 1, figs. 34 and 35.

c. 1510. "Jagellonicus" Globe. Reproduced in Stevenson, *Terrestrial and Celestial Globes,* vol. 1, fig. 37.

* c. 1510. Henricus Loritus Glareanus World Map. Switzerland. Reproduced in plate 10.

* c. 1510–1520. Henricus Loritus Glareanus Map of the Pacific. Switzerland. Reproduced in plate 11.

c. 1510–1520. Henricus Loritus Glareanus Map of the Southern Hemisphere. Switzerland. Reproduced in Wroth, "The Early Cartography of the Pacific," plate 6.

c. 1510–1520. Henricus Loritus Glareanus Map of the Northern Hemisphere. Switzerland. Reproduced in Wroth, *The Voyages of Giovanni da Verrazzano,* plate 10.

1511. Vesconte de Maggiolo World Map. Naples. Reproduced in Nebenzahl, *Atlas of Columbus and the Great Discoveries,* pp. 58–59.

1511. "Peter Martyr" Map of the Indies. (Anonymous map attributed to Andrea Morales). Seville. Reproduced in Nebenzahl, *Atlas of Columbus and the Great Discoveries,* p. 61.

1511. Bernard Sylvanus World Map. Venice. Reproduced in Shirley, *The Mapping of the World*, p. 36.

1512. Johannes de Stobnicza Map of the Pacific. Kraków. Reproduced in Shirley, *The Mapping of the World*, p. 37.

1513. Piri Re'is Chart of the Ocean Sea. Gallipoli. Reproduced in Nebenzahl, *Atlas of Columbus and the Great Discoveries*, p. 63.

1513. Martin Waldseemüller *Terre Nove*. Strassburg. Reproduced in Nebenzahl, *Atlas of Columbus and the Great Discoveries*, pp. 64–65.

* c. 1513. Anonymous Portuguese map of the Pacific. Reproduced in plate 25, from Winsor, *Narrative and Critical History of America*, vol. 2, p. 440.

1514. Cornelius Aurelius World Map. Leiden. Reproduced in Shirley, *The Mapping of the World*, p. 42.

c. 1514. "Leonardo da Vinci" World Map. Reproduced in Nordenskiöld, *A Facsimile-Atlas to the Early History of Cartography*, p. 77.

c. 1514. Louis Boulengier Globe Gores. Lyons. Reproduced in Shirley, *The Mapping of the World*, p. 43.

1515. Johann Schöner Globe. Germany. Reproduced in Wroth, *The Voyages of Giovanni da Verrazzano*, plate 13.

1515. Gregor Reisch World Map. Strassburg. Reproduced in Shirley, *The Mapping of the World*, p. 46.

c. 1515. "Paris (Green, Quirini)" Globe. Reproduced in Wroth, *The Voyages of Giovanni da Verrazzano*, plate 14.

1516. Martin Waldseemüller *Carta Marina*. Strassburg. Reproduced in Shirley, *The Mapping of the World*, pp. 48–49.

* c. 1518. "Kunstmann IV" Map of the Western Hemisphere. (Part of a larger anonymous Portuguese world map generally attributed to Jorge Reinel and reproduced in its entirety in Levillier, *America la Bien Llamada*, pp. 142–143.) Reproduced in plate 31.

c. 1518. "Liechtenstein (Nordenskiöld)" Globe Gores. Reproduced in Wroth, *The Voyages of Giovanni da Verrazzano*, plate 15.

c. 1519. Lopo Homem and Pedro Reinel Map of the East

Indies. Portugal. Reproduced in Nebenzahl, *Atlas of Columbus and the Great Discoveries*, pp. 70–71.

c. 1519. Alonso Álvarez de Pineda Map of the Gulf of Mexico. Reproduced in Wroth, *The Voyages of Giovanni da Verrazzano*, plate 16.

* 1520. Johann Schöner Globe. Germany. Reproduced in plate 9.

1520. Peter Apian World Map. Vienna. Reproduced in Shirley, *The Mapping of the World*, p. 52.

1525. "Castiglioni" World Map. Reproduced in Cortesão, "Notes on the Castiglioni Planisphere."

* 1526. Juan Vespucci World Map. Seville. Reproduced in plates 15 and 17.

* c. 1526. "Salviati" World Map. Reproduced in plates 16 and 18.

c. 1527. Franciscus Monachus Map of the Western Hemisphere. Antwerp. Reproduced in Shirley, *The Mapping of the World*, p. 61.

1528. Pietro Coppo World Map. Venice. Reproduced in Shirley, *The Mapping of the World*, p. 65.

* c. 1528. "Gilt (De Bure)" Globe. Reproduced in plates 19 and 20.

1529. Diego Ribero World Map. Seville. Reproduced in Nebenzahl, *Atlas of Columbus and the Great Discoveries*, pp. 93–95.

c. 1530. "Nancy" Globe. Reproduced in Nordenskiöld, *Periplus*, p. 159.

1531. Oronce Finé Double-Cordiform World Map. Paris. Reproduced in Shirley, *The Mapping of the World*, pp. 72–73.

1532. Simon Grynaeus World Map. Basle. Reproduced in Nordenskiöld, *A Facsimile-Atlas to the Early History of Cartography*, plate 42.

1534. Joachim von Watte World Map. Zurich. Reproduced in Shirley, *The Mapping of the World*, p. 78.

* 1534. Oronce Finé Cordiform World Map. Paris. Reproduced in plate 1.

c. 1535. Georg Hartmann Globe Gores. Nuremberg. Reproduced in Shirley, *The Mapping of the World*, p. 82.

c. 1535. Anonymous World Map. Reproduced in Skelton, *Explorers' Maps*, pp. 66–67.

* c. 1535. Anonymous Globe Gores. Nuremberg. Reproduced in plate 12.

1536. Caspar Vopell Globe Gores. Cologne. Reproduced in Shirley, *The Mapping of the World*, p. 83.

1538. Gerardus Mercator Double-Cordiform World Map. Louvain. Reproduced in Nordenskiöld, *A Facsimile-Atlas to the Early History of Cartography*, plate 43.

1542. Battista Agnese Oval World Map. Venice. Reproduced in Nebenzahl, *Atlas of Columbus and the Great Discoveries*, pp. 102–103.

1542. Battista Agnese World Map. Venice. Reproduced in Nebenzahl, *Atlas of Columbus and the Great Discoveries*, pp. 100–101.

* 1544. Sebastian Münster Map of the New Islands. Basle. Reproduced in plate 23.

1544. Sebastian Cabot World Map. Antwerp. Reproduced in Nebenzahl, *Atlas of Columbus and the Great Discoveries*, pp. 106–107.

1545. Caspar Vopell World Map. (See 1558 Vavassore and 1570 Van den Putte copies).

1546. Johann Honter World Map. Zurich. Reproduced in Shirley, *The Mapping of the World*, p. 98.

1546. Giacomo Gastaldi World Map. Venice. Reproduced in Shirley, *The Mapping of the World*, pp. 96–97.

1548. Giacomo Gastaldi World Map. Venice. Reproduced in Shirley, *The Mapping of the World*, p. 100.

* c. 1550. Giacomo Gastaldi and Matteo Pagano World Map. Venice. Reproduced in plate 27.

1552. François Demongenet Globe Gores. Reproduced in Shirley, *The Mapping of the World*, p. 106.

1554. Michele Tramezzino World Map. Italy. Reproduced in Shirley, *The Mapping of the World*, pp. 110–111.

1555. Giacomo Gastaldi and Gerard de Jode World Map. Antwerp. Reproduced in Shirley, *The Mapping of the World*, pp. 114–115.

c. 1555. Giorgio Calapoda World Map. Italy. Reproduced in Shirley, *The Mapping of the World*, p. 112.

* 1558. Giovanni Vavassore's Copy of the 1545 Caspar Vopell World Map. Venice. Reproduced in plates 28 and 29.

c. 1559. Haggi Ahmed World Map. Venice. Reproduced in Shirley, *The Mapping of the World*, p. 118.

1560. Paolo Forlani World Map. Venice. Reproduced in Shirley, *The Mapping of the World*, p. 122.

c. 1560. François Demongenet Globe Gores. Reproduced in Shirley, *The Mapping of the World*, p. 120.

1561. Girolamo Ruscelli World Map. Venice. Reproduced in Shirley, *The Mapping of the World*, p. 127.

c. 1561. Giacomo Gastaldi World Map. Venice. Reproduced in Shirley, *The Mapping of the World*, pp. 124–125.

1562. Paolo Forlani World Map. Venice. Reproduced in Shirley, *The Mapping of the World*, pp. 128–129.

1565. Paolo Forlani World Map. Venice. Reproduced in Shirley, *The Mapping of the World*, p. 134.

1566. Bolognino Zaltieri Map of North America. Italy. Reproduced in Nordenskiöld, *A Facsimile-Atlas to the Early History of Cartography*, p. 129.

1566. Giovanni Cimerlino World Map. Italy. Reproduced in Shirley, *The Mapping of the World*, p. 136.

1567. Giovanni Francesco Camocio World Map. Venice. Reproduced in Shirley, *The Mapping of the World*, p. 138.

1569. Gerardus Mercator World Map. Duisburg. Reproduced in Nebenzahl, *Atlas of Columbus and the Great Discoveries*, pp. 128–129.

1570. Paolo Forlani and Claudio Duchetti World Map. Venice. Reproduced in Shirley, *The Mapping of the World*, pp. xxvi–xxvii.

1570. Bernard Van den Putte Copy of the 1545 Caspar Vopell World Map. Antwerp. Reproduced in Shirley, *The Mapping of the World*, pp. 148–149.

1571. Benito Arias Map of the Western Hemisphere. Antwerp. Reproduced in Shirley, *The Mapping of the World*, p. 150.

1572. Tommaso Porcacchi World Map. Venice. Reproduced in Shirley, *The Mapping of the World*, p. 151.

1574. Georg Braun World Map. Cologne. Reproduced in Shirley, *The Mapping of the World*, pp. 154–155.

1579. Mario Cartaro Astronomical Diagrams. Rome. Reproduced in Shirley, *The Mapping of the World*, p. 163.

c. 1587. Giacomo Franco Cordiform World Map. Italy. Reproduced in Shirley, *The Mapping of the World*, p. 174.

1590. Joannes Myritius World Map. Ingolstadt. Reproduced in Shirley, *The Mapping of the World*, pp. 194–195.

c. 1590. Sigurdur Stefánsson Map of the North Atlantic. Iceland. Reproduced in Skelton et al., *The Vinland Map and the Tartar Relation*, plate 17.

* 1599. Edward Wright World Chart on Mercator Projection. London. Reproduced in plate 24.

1605. Hans Poulson Resen Map of the North Atlantic. Denmark. Reproduced in Skelton et al., *The Vinland Map and the Tartar Relation*, plate 19.

1710–1711. Fedor Beiton Iakutsk Map. Russia. Reproduced in Fisher, *Bering's Voyages*, p. 41.

c. 1710–1715. Ivan L'vov Anadyrsk Map. Russia. Reproduced in Fisher, *Bering's Voyages*, p. 46.

1712–1714. Semen Remezov Map of Kamchatka. Russia. Reproduced in Fisher, *Bering's Voyages*, p. 39.

1750. Joseph Nicolas Delisle Map of the North Pacific. Paris. Reproduced in Müller, *Bering's Voyages*, map 15.

* c. 1751. Isaac Tirion Map of the Arctic Pole. Amsterdam. Reproduced in plate 26, from Portinaro and Knirsch, *The Cartography of North America*, p. 239.

1758. Gerhard Friedrich Müller Map of the Russian Discoveries. St. Petersburg. Reproduced in Cook, *Flood Tide of Empire*, illus. 10.

1772. Didier Robert de Vaugondy Map of North and West America. France. Reproduced in Portinaro and Knirsch, *The Cartography of North America*, pp. 284–285.

BIBLIOGRAPHY

Anderson, Charles L. G. *Life and Letters of Vasco Núñez de Balboa.* Westport, Conn.: Greenwood, 1941.

Archer, Christon I. "Russians, Indians, and Passages: Spanish Voyages to Alaska in the Eighteenth Century." In *Exploration in Alaska*, edited by Antoinette Shalkop, pp. 129–143. Anchorage: Cook Inlet Historical Society, 1980.

Arciniegas, Germán. *Amerigo and the New World.* New York: Alfred A. Knopf, 1955.

Bagrow, Leo. "The First Russian Maps of Siberia and their Influence on the West-European Cartography of N. E. Asia." *Imago Mundi* 9 (1952):83–93.

Bailey, James. *The God-Kings and the Titans.* New York: St. Martin's Press, 1973.

Barratt, Glynn. *Russia in Pacific Waters 1715–1825.* Vancouver: University of British Columbia Press, 1981.

Beaglehole, J. C., ed. *The Voyage of the Resolution and Discovery 1776–1780.* London: Cambridge University Press, 1967.

Beals, Herbert K. *For Honor and Country.* Portland: Oregon Historical Society, 1985.

———. *Juan Pérez on the Northwest Coast.* Portland: Oregon Historical Society, 1989.

Bigelow, John. "The So-Called Bartholomew Columbus Map of 1506." *Geographical Review* 25 (1935):643–656.

Bloomer, Carolyn M. *Principles of Visual Perception.* New York: Van Nostrand Reinhold, 1976.

Bloor, David. "Polyhedra and the Abominations of Leviticus: Cognitive Styles in Mathematics." In *Essays in the Sociology of Perception*, edited by Mary Douglas, pp. 191–218. London: Routledge & Kegan Paul, 1982.

Boland, Charles M. *They All Discovered America*. 1961. Reprint. New York: Pocket Books, 1963.

Boorstin, Daniel J. *The Discoverers*. New York: Random House, 1983.

Breitfuss, L. "Early Maps of North-Eastern Asia and of the Lands around the North Pacific." *Imago Mundi* 3 (1939): 87–99.

Brøgger, A. W., and Haakon Shetelig. *The Viking Ships*. Oslo: Dreyers, 1951.

Butterfield, C. W. *History of the Discovery of the Northwest by John Nicolet in 1634*. 1881. Reprint. Port Washington, N.Y.: Kennikat Press, 1969.

Campbell, Tony. "Portolan Charts from the Late Thirteenth Century to 1500." In *The History of Cartography*, edited by J. B. Harley and David Woodward, vol. 1, pp. 371–463. Chicago: University of Chicago Press, 1987.

Columbus, Ferdinand. *The Life of the Admiral Christopher Columbus*. Before 1539. Translated and edited by Benjamin Keen, 1959. Reprint. New Brunswick, N.J.: Rutgers University Press, 1992.

Cook, Warren L. *Flood Tide of Empire*. New Haven: Yale University Press, 1973.

Cortesão, Armando. "The North Atlantic Nautical Chart of 1424." *Imago Mundi* 10 (1953):1–13.

———. "Notes on the Castiglioni Planisphere." *Imago Mundi* 11 (1954):53–55.

———. *History of Portuguese Cartography*. Coimbra, Portugal: Junta de Investigações do Ultramar, 1969.

Davies, Arthur. "The 'English' Coasts on the Map of Juan de la Cosa." *Imago Mundi* 13 (1956):26–29.

———. "The Date of Juan de la Cosa's World Map and Its Implications for American Discovery." *Geographical Journal* 142 (1976):111–116.

Day, A. Grove. *Coronado's Quest*. Berkeley: University of California Press, 1940.

Dunn, Oliver, and James E. Kelley, eds. *The Diario of Christopher Columbus's First Voyage to America 1492–1493*. Norman: University of Oklahoma Press, 1989.

Enterline, J. "The Southern Continent and the False Strait of Magellan." *Imago Mundi* 26 (1972):48–58.

Fell, Barry. *America B.C.* Rev. ed. New York: Pocket Books, 1989.

———. *Saga America*. New York: Times Books, 1980.

Fernández-Armesto, Felipe, ed. *The Times Atlas of World Exploration*. New York: HarperCollins, 1991.

Fisher, Raymond H. *Bering's Voyages*. Seattle: University of Washington Press, 1977.

———. *The Voyage of Semen Dezhnev in 1648*. London: Hakluyt Society (2nd series, vol. 159), 1981.

Fleck, Ludwik. *Genesis and Development of A Scientific Fact*. 1935. Reprint. Chicago: University of Chicago Press, 1981.

Forster, Johann R. *History of the Voyages and Discoveries Made in the North*. Dublin: 1786.

Frenkel-Brunswik, Else. "Intolerance of Ambiguity as an Emotional and Perceptual Personality Variable." *Journal of Personality* 18 (1949):108–143.

Fuson, Robert H., trans. *The Log of Christopher Columbus*. Camden, Maine: International Marine Publishing Co., 1987.

Ganong, W. F. *Crucial Maps in the Early Cartography and Place-Nomenclature of the Atlantic Coast of Canada*. Toronto: University of Toronto Press, 1964.

Garfinkel, Harold. *Studies in Ethnomethodology*. Englewood Cliffs, N.J.: Prentice-Hall, 1967.

Golder, Frank A. *Russian Expansion on the Pacific, 1641–1850*. Cleveland, Ohio: Arthur H. Clark, 1914.

———. *Bering's Voyages*. 2 vols. New York: American Geographical Society, 1922.

Gordon, Cyrus H. *Before Columbus*. New York: Crown, 1971.

Harrisse, Henry. *The Discovery of North America*. 1892. Reprint. Amsterdam: N. Israel, 1969.

Heawood, Edward. "Glareanus: His Geography and Maps." *Geographical Journal* 25 (1905):647–654.

Helms, Mary W. *Ulysses' Sail*. Princeton: Princeton University Press, 1988.

Hirsch, Rudolf. "Printed Reports on the Early Discoveries and Their Reception." In *First Images of America*, edited by Fredi Chiappelli, vol. 2, pp. 537–552. Berkeley: University of California Press, 1976.

Hodge, Frederick W., ed. *Spanish Explorers in the Southern United States 1528–1543*. 1907. Reprint. New York: Barnes & Noble, 1946.

Hoffman, Bernard G. *Cabot to Cartier*. Toronto: University of Toronto Press, 1961.

Howse, Derek. *Greenwich Time and the Discovery of the Longitude*. Oxford: Oxford University Press, 1980.

Ingstad, Helge. *Westward to Vinland*. New York: St. Martin's Press, 1969.

Irwin, Constance. *Fair Gods and Stone Faces*. New York: St. Martin's Press, 1963.

Jane, Cecil, ed. *The Four Voyages of Columbus*. 2 vols. 1930, 1933. Reprint. New York: Dover, 1988.

Jones, Gwyn. "The Western Voyages and the Vinland Map." In *Proceedings of the Vinland Map Conference*, edited by Wilcomb E. Washburn, pp. 119–127. Chicago: University of Chicago Press, 1971.

———. *The Norse Atlantic Saga*. 2nd ed. Oxford: Oxford University Press, 1986.

Kessler, Suzanne J., and Wendy McKenna. *Gender*. New York: John Wiley, 1978.

Kimble, George H. T. "Introduction" to Duarte Pacheco Pereira, *Esmeraldo de Situ Orbis*. London: Hakluyt Society (2nd series, no. 79), 1937.

Koenigsberg, Richard A. *Symbiosis and Separation*. New York: Library of Art and Social Science, 1989.

Kohl, Johann G. "Asia and America." *Proceedings of the American Antiquarian Society* 21 (1911):284–338.

Kuhn, Thomas S. *The Structure of Scientific Revolutions.* 2nd ed. Chicago: University of Chicago Press, 1970.

Kunstmann, Friedrich, et al. *Atlas zur Entdeckungsgeschichte Amerikas.* Munich: 1859.

Lakatos, Imre. *Proofs and Refutations.* Cambridge: Cambridge University Press, 1976.

Las Casas, Bartolomé de. *History of the Indies.* Edited by Andrée Collard. c. 1550–1565. Reprint. New York: Harper Torchbooks, 1971.

Levillier, Roberto. *America la Bien Llamada.* Buenos Aires: Guillermo Kraft, 1948.

————. "New Light on Vespucci's Third Voyage: Evidence of His Route and Landfalls." *Imago Mundi* 11 (1954):37–46.

Magnusson, Magnus, and Hermann Pálsson, trans. *The Vinland Sagas.* Harmondsworth: Penguin, 1965.

Mahler, Margaret S., et al. *The Psychological Birth of the Human Infant.* New York: Basic Books, 1975.

Mallery, Arlington, and Mary R. Harrison. *The Rediscovery of Lost America.* 1951. Reprint. New York: E. P. Dutton, 1979.

Mannheim, Karl. *Ideology and Utopia.* New York: Harvest, 1936.

Martyr, Peter. *The Decades of the New World or West India.* In *The First Three English Books on America,* edited by Edward Arber, pp. 63–200. 1511–1555. Reprint. Birmingham: 1885.

Morison, Samuel E. *Portuguese Voyages to America in the Fifteenth Century.* Cambridge: Harvard University Press, 1940.

————. *The European Discovery of America—The Northern Voyages* A.D. *500–1600.* New York: Oxford University Press, 1971.

————. *The European Discovery of America—The Southern Voyages* A.D. *1492–1616.* New York: Oxford University Press, 1974.

Mourelle, Francisco A. *Voyage of the Sonora in the Second Bucareli Expedition.* 1781. Reprint. San Francisco: Thomas C. Russell, 1920.

Müller, Gerhard F. *Bering's Voyages—the Reports from Russia.* 1758. Reprint. Fairbanks: University of Alaska Press, 1986.

Nebenzahl, Kenneth. *Atlas of Columbus and the Great Discoveries.* Chicago: Rand McNally, 1990.

Newby, Eric. *The World Atlas of Exploration*. 1975. Reprint. New York: Crescent, 1985.

Newton, Arthur P. "Asia or Mundus Novus?" In *The Great Age of Discovery*, edited by Arthur P. Newton, pp. 104–128. 1932. Reprint. New York: Burt Franklin, 1970.

Nordenskiöld, A. E. *Facsimile-Atlas to the Early History of Cartography*. 1889. Reprint. New York: Dover, 1973.

———. *Periplus*. Stockholm: P. A. Norstedt, 1897.

Nunn, George E. *The Geographical Conceptions of Columbus*. New York: American Geographical Society, 1924.

———. "The Three Maplets Attributed to Bartholomew Columbus." *Imago Mundi* 9 (1952):12–22.

O'Gorman, Edmundo. *The Invention of America*. Bloomington: Indiana University Press, 1961.

Pacheco Pereira, Duarte. *Esmeraldo de Situ Orbis*. Translated and edited by George H. T. Kimble. 1505–1508. Reprint. London: Hakluyt Society (2nd series, no. 79), 1937.

Parry, John H. *The Discovery of South America*. New York: Taplinger Publishing Co., 1979.

Parry, John H., and Robert G. Keith, eds. *New Iberian World*. 5 vols. New York: Times Books and Hector & Rose, 1984.

Pennington, Piers. *The Great Explorers*. 1979. Reprint. London: Bloomsbury, 1989.

Pohl, Frederick J. *Amerigo Vespucci—Pilot Major*. New York: Columbia University Press, 1944.

Polo, Marco. *The Travels of Marco Polo the Venetian*. Edited by Manuel Komroff. c. 1300. Reprint. New York: Boni & Liveright, 1926.

Portinaro, Pierluigi, and Franco Knirsch. *The Cartography of North America 1500–1800*. New York: Crescent, 1987.

Quinn, David B. "The Argument for the English Discovery of America between 1480 and 1494." *Geographical Journal* 127 (1961):277–285.

———. "John Day and Columbus." *Geographical Journal* 133 (1967):205–209.

———. *England and the Discovery of America, 1481–1620*. New York: Alfred A. Knopf, 1974.

————. "New Geographical Horizons." In *First Images of America*, edited by Fredi Chiappelli, vol. 2, pp. 635–658. Berkeley: University of California Press, 1976.

Quinn, David B., ed. *New American World*. 5 vols. New York: Arno Press & Hector Bye, 1979.

Ramsay, Raymond H. *No Longer on the Map*. New York: Ballantine, 1972.

Randles, W.G.L. "The Evaluation of Columbus's 'India' Project by Portuguese and Spanish Cosmographers in the Light of the Geographical Science of the Period." *Imago Mundi* 42 (1990): 50–64.

Robertson, William. *The History of America*. London: Strahan, Cadell, and Balfour, 1777.

Ruddock, Alwyn A. "Columbus and Iceland: New Light on an Old Problem." *Geographical Journal* 136 (1970):177–189.

Sale, Kirkpatrick. *The Conquest of Paradise*. New York: Alfred A. Knopf, 1990.

Sauer, Carl O. *Sixteenth Century North America*. Berkeley: University of California Press, 1971.

Shibutani, Tamotsu. "Reference Groups as Perspectives." *American Journal of Sociology* 60 (1955):562–569.

Shirley, Rodney W. *The Mapping of the World*. 1984. Reprint. London: The Holland Press, 1987.

Silverman, Lloyd, et al. *The Search for Oneness*. New York: International Universities Press, 1982.

Skelton, R. A. *Explorers' Maps*. London: Routledge & Kegan Paul, 1958.

————. "The Vinland Map." In R. A. Skelton et al., *The Vinland Map and the Tartar Relation*, pp. 109–240. New Haven: Yale University Press, 1965.

Skelton, R. A., et al. *The Vinland Map and the Tartar Relation*. New Haven: Yale University Press, 1965.

Stefansson, Vilhjalmur. *Ultima Thule*. New York: Macmillan, 1940.

Steller, Georg W. *Journal of a Voyage with Bering 1741–1742*. 1743. Reprint. Stanford: Stanford University Press, 1988.

Stevenson, Edward L. *Terrestrial and Celestial Globes*. 2 vols. New Haven: Yale University Press, 1921.

Taylor, E.G.R. "Idée Fixe: The Mind of Christopher Columbus." *Hispanic American Historical Review* 11 (1931):289–301.

Thacher, John B. *Christopher Columbus*. 3 vols. New York: G. P. Putnam's Sons, 1903.

Tompkins, Stuart R. "After Bering: Mapping the North Pacific." *British Columbia Historical Quarterly* 19 (1955):1–55.

Tompkins, Stuart R., and Max L. Moorhead. "Russia's Approach to America." *British Columbia Historical Quarterly* 13 (1949): 55–66, 231–255.

True, David O. "Some Early Maps Relating to Florida." *Imago Mundi* 11 (1954):73–84.

Van Sertima, Ivan. *They Came Before Columbus*. New York: Random House, 1976.

Vigneras, Louis-André. "New Light on the 1497 Cabot Voyage to America." *Hispanic American Historical Review* 36 (1956): 503–509.

———. "The Cape Breton Landfall: 1494 or 1497." *Canadian Historical Review* 38 (1957):219–228.

———. Review of *The Vinland Map and the Tartar Relation*, by R. A. Skelton, et al. *English Historical Review* 83 (1968): 117–119.

———. *The Discovery of South America and the Andalusian Voyages*. Chicago: University of Chicago Press, 1976.

Von Wuthenau, Alexander. *Unexpected Faces in Ancient America 1500 B.C.–1500 A.D.* New York: Crown, 1975.

Wagner, Henry R. *Spanish Voyages to the Northwest Coast of America in the Sixteenth Century*. San Francisco: California Historical Society, 1929.

———. *The Cartography of the Northwest Coast of America to the Year 1800*. 2 vols. Berkeley: University of California Press, 1937.

Wahlgren, Erik. *The Kensington Stone—A Mystery Solved*. Madison: University of Wisconsin Press, 1958.

Waldseemüller, Martin. *Cosmographiae Introductio*. 1507. Reprint. Ann Arbor, Mich.: University Microfilms Inc., 1966.

Wallace, Birgitta L. "The L'Anse aux Meadows Site." In Gwyn

Jones, *The Norse Atlantic Saga*, pp. 285–304. 2nd ed. Oxford: Oxford University Press, 1986.

Washburn, Wilcomb E. "Japan on Early European Maps." *Pacific Historical Review* 21 (1952):221–236.

———. "The Meaning of 'Discovery' in the Fifteenth and Sixteenth Centuries." *American Historical Review* 68 (1962):1–21.

Washburn, Wilcomb E., ed. *Proceedings of the Vinland Map Conference*. Chicago: University of Chicago Press, 1971.

Weinstein, Donald. *Ambassador from Venice*. Minneapolis: University of Minnesota Press, 1960.

Williams, Glyndwr. "Alaska Revealed: Cook's Explorations in 1778." In *Exploration in Alaska*, edited by Antoinette Shalkop, pp. 69–87. Anchorage: Cook Inlet Historical Society, 1980.

Williamson, James A. *The Cabot Voyages and Bristol Discovery under Henry VII*. Cambridge: Cambridge University Press, 1962.

Winnicott, D. W. "Transitional Objects and Transitional Phenomena." In *Playing and Reality*, pp. 1–25. 1953. Reprint. London: Tavistock, 1971.

Winsor, Justin. *Narrative and Critical History of America*. 8 vols. Boston: Houghton Mifflin & Co., 1886.

Wroth, Lawrence C. "The Early Cartography of the Pacific." *Papers of the Bibliographical Society of America* 38 (1944):87–268.

———. *The Voyages of Giovanni da Verrazzano 1524–1528*. New Haven: Yale University Press, 1970.

Zerubavel, Eviatar. *Patterns of Time in Hospital Life*. Chicago: University of Chicago Press, 1979.

———. "If Simmel Were a Fieldworker: On Formal Sociological Theory and Analytical Field Research." *Symbolic Interaction* 3 (1980), 2:25–33.

———. "The Standardization of Time: A Sociohistorical Perspective." *American Journal of Sociology* 88 (1982):1–23.

———. *The Seven-Day Circle*. New York: Free Press, 1985.

———. *The Fine Line*. New York: Free Press, 1991.

Zerubavel, Yael. "The Politics of Interpretation: Tel Hai in Israel's Collective Memory." *AJS Review* 16 (1991):133–159.

———. *The Politics of Commemoration*. Forthcoming.

Zweig, Stefan. *Amerigo*. New York: Viking, 1942.

INDEX